The Spirit
of the Hive

The
Spirit
of the
Hive

The Mechanisms of Social Evolution

Robert E. Page, Jr.

Harvard University Press

Cambridge, Massachusetts

London, England

2013

Library of Congress Cataloging-in-Publication Data
Page, Robert E.
The spirit of the hive : the mechanisms of social evolution / Robert E. Page, Jr.
pages   cm
Includes bibliographical references and index.
ISBN 978-0-674-07302-9 (alkaline paper)
1. Honeybee—Evolution.   2. Honeybee—Behavior   3. Beehives.
4. Pollen.   I. Title.
QL568.A6P24   2013
595.79'9—dc23   2012044923

To Harry Laidlaw (1907–2003)
My good friend, teacher, and mentor

Contents

Foreword

The late Larry Slobodkin, the distinguished evolutionary geneticist and theoretician, once said, "Nature defeats theory." I was frequently reminded of this wise revelation when I read *The Spirit of the Hive*. This book provides an avalanche of hard facts. Yes, it contains models too, but models that are based on solid scientific facts to serve as tools to design new experiments and to investigate unknown perspectives of the regulatory networks that constitute the foundation of division of labor in honey bee societies.

This concisely written book by Robert Page opens a window for a deeper view of the genetic and physiological architecture of the life of the honey bee and its society. Whereas other recently published books on honey bees inform us about the ways an insect society of about 40,000 individuals organizes division of labor, how its members communicate with one another, and how they make collective decisions, this challenging book by Robert Page explores the genetic and physiological mechanisms on the individual and colony levels that underlie the evolution and workings of honey bee society. It presents new, exciting insights concerning the so-called building blocks, the raw material, on which natural selection operates during the evolution of eusocial behavior. Although, as Robert Page states, the word *superorganism* hardly appears on the pages of his book, his work is deeply intertwined with the superorganism concept. In fact, this book offers the most thorough analysis of the physiological mechanisms and genetic substrate that underlie the emergence of colony traits, and it provides the best experimental evidence for colony-level selection, which of course, does not contradict inclusive-fitness theory, as other authors occasionally have erroneously claimed. The intricate social interactions within honey bee society deeply affect development and ontogeny of all its members and lead to the manifestation of robust colony traits.

This is an important scientific book. It covers the lifetime work of Robert Page and his impressive, international group of collaborators, mostly his former doctoral and postdoctoral students. The results of these far-reaching experimental studies, obtained during the past 30 years, have been published in many first-rate journals, such as *Nature, Science, Proceedings of the National Academy of Sciences of the United States of America, Genetics,* and *Insect Molecular Biology,* but this volume presents the first comprehensive synthesis of this important work. The book is written for scientists, but it is also very suitable for graduate student courses in biology. Each chapter offers plenty of material for discussion with advanced students.

The book is exemplary in its focus on collaborative research and in using interdisciplinary approaches to make new progress in understanding complex adaptive systems. Although the statement appears obvious, it cannot be emphasized enough that the scientific enterprise is conducted by people, and success in this endeavor very much depends on the quality of human interactions, be they competitive or collaborative.

Bert Hölldobler

Preface

In 2009–2010, I spent 10 months as a fellow at the Wissenschaftskolleg zu Berlin (Institute for Advanced Studies) in Berlin, Germany. I was part of a study group that investigated developmental evolution in social insects. Although writing a book is the normal activity of WIKO (the familiar name for the Wissenschaftskolleg) fellows, I arrived with the explicit and frequently announced plan that I would not write one. In the biological sciences, research papers are the coin of the realm, and there is little to gain with respect to advancing one's career by writing books. However, my good friend, colleague, and author of the foreword of this book, Bert Hölldobler, had encouraged me for several years to write one. He thought that the body of work produced by my lab and colleagues over the past 30 years or more was too widely scattered—across many research papers in journals spanning multiple disciplines—and should be consolidated into a single, coherent volume. But I resisted.

While I was at the WIKO, I was invited to give the annual Ernst Mayr Lecture to the Berlin-Brandenburg Academy of Science. The title of my talk was "The Spirit of the Hive," and its theme was developmental evolution. The academy publishes a small booklet for each lecture. Afterward, Bert suggested that I take the booklet derived from the lecture as a starting point and write the book. I gave in and began writing in January 2010, and by the time I left WIKO in mid-July, I had a first draft. Completion of the book has been a much more laborious process requiring frequent revisits to Berlin.

My approach in this book is different from that of most books on social insect behavior and evolution. I think that this is because of my background. I was not trained as an ecologist or behavioral ecologist. I am an entomologist trained in traditional breeding genetics, evolutionary population genetics, and mechanisms of behavior. As a consequence, after this preface, you will not see the words *kin selection* or *superorganism,* although they are interesting and important topics. This book is not about them. It is also not about behavioral ecology; therefore, it contains very little about how the external environment shapes the behavior of individuals or the social organization of colonies. Behavioral ecology is more the domain of the very interesting book by Tom Seeley, *The Wisdom of the Hive.* And although our titles are very similar, our books are very different. This book is about how social organization is linked vertically from genes to societies by mechanisms that evolve and operate within and between biological levels. I map out the architecture of traits and mechanisms and show how selection has altered them.

The main theme of this book is "the spirit of the hive." I explain it in Chapter 2 as the stimulus-response relationships of individuals with their environment. The book is centered on how those relationships shape individual and social behavior. I start with a "stone soup model" of division of labor and add complexity throughout the book. I present my view on models and modeling and develop heuristic tools along the way in an attempt to bring understanding to the mechanisms of social organization and their evolution and to generate new questions—a second theme.

I have not attempted to cite all the excellent work of all the researchers in the field. Instead, I support the content of each chapter with suggested readings listed at the end. I want this to be a book that is read for the story it tells, not as a collection of scientific references. This book is a tale of more than 30 years of research by my students, postdocs, colleagues, and me. The story is from my perspective, although the research ideas, development, and advancements were truly collaborative and shared by all. But I accept singular responsibility for the errors.

— 1 —

Darwin's Dilemma and
the Spirit of the Hive

Social insects have fascinated natural historians and philosophers since Aristotle and continue to fascinate us today with their self-sacrificing altruism, complex nest architecture, untiring industry, and division of labor. They presented Charles Darwin with special difficulties for his fledgling theory of evolution by natural selection. How can sterile castes, such as worker honey bees, wasps, and ants, evolve when they don't normally reproduce? The existence of sterile castes seems to be in direct opposition to a theory that requires differential survival and reproductive success. Darwin considered an even bigger difficulty to be the observation that the reproductive individuals in colonies are often anatomically differentiated from the sterile workers, showing adaptation of a sterile caste. However, he considered the biggest difficulty to be the anatomical differentiation within the worker caste that is dramatically demonstrated in many species of ants. Darwin waved his arms and invoked selection on families as an explanation, an explanation later shown to be too simplistic.

Social insects provided additional difficulties for Darwin when he considered the architecture of the honey bee nest (Figure 1.1). Darwin asked a Cambridge mathematician to study the comb of the bee from an engineering perspective of strength and economy, concluding, "For the comb of the hive bee, as far as we can see, is absolutely perfect in economizing labour and wax." How could the wax combs be built with such precision to maximize the strength of the comb and at the same

1

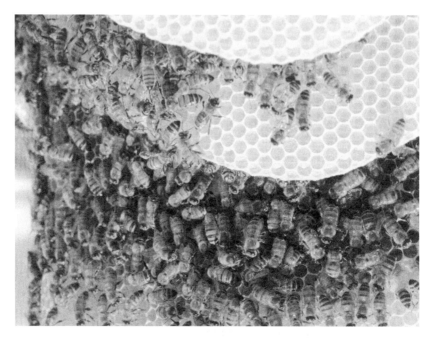

Figure 1.1. A colony of honey bees engaged in the construction of wax comb. Photo by Jacob Sahertian.

time save costly building materials? And, as he pointed out, "this is effected by a crowd of bees working in a dark hive" (Darwin 1998, pp. 348–349). How could they achieve this architectural masterpiece with instincts alone, working without any central control of construction tasks? Darwin experimented with honey bees and demonstrated to his satisfaction that bees could construct combs using just their instincts and local information regarding cell construction, thereby solving his dilemma of perfection and instincts.

The Nobel Laureate poet, playwright, and author Maurice Maeterlinck was also fascinated by social insects. In his wonderfully romantic book *The Life of the Bee,* originally published in 1901, he noted that there was no central control of cooperative behavior, thought by many to be the domain of the queen, and stated, "She is not the queen in the sense in which men use the word. She issues no orders; she obeys, as meekly as the humblest of her subjects, the masked power, sovereignly wise, that for the present, and till we attempt to locate it, we will term the 'spirit of the hive.'" (pp. 38–39; English translation 1903). Here he resorted to a mystical vitalism to explain how colonies full of individuals working in the dark organize into a cooperative whole, and he left it for someone else to identify the "spirit of the hive" and where it resides.

How do insect societies evolve complex social organization? There is a hierarchy of organizational levels from genes to the society, but there is no centralized control of behavior, and there is no social genome controlling the society on which natural selection can act. Yet it happens. Insects display the most complex and fascinating social organization known, enough to capture the minds and fantasies of both Darwin and Maeterlinck. In the following chapters, I will define the genetic, physiological, and behavioral mechanisms behind the mystical "spirit of the hive" of Maeterlinck. In Chapter 2, I will show how division of labor among worker honey bees arises from the simple result that bees, like all other animals, respond to stimuli in their environments and, as a consequence, change the stimulus environment. In Chapter 3, I show how genetic variation within colonies resulting from the polyandrous mating behavior of queens—queens mate with many males—contributes to variation in the stimulus-response relationships of workers within a nest and contributes to their social organization. Thus there are consequences of polyandry. In Chapter 4, I propose that polyandry may have evolved in response to its effects on within-colony genetic variation and social organization and dynamics, although there are many competing hypotheses.

Chapters 5 and 6 focus on results of a selective breeding program that has continued for more than 20 years, designed to study the effects of selection on a single colony-level trait, the amount of surplus stored pollen. This trait is a consequence of the complex interactions of thousands of individuals who share the nest. I present the phenotypic architecture— the correlated changes in many different behavioral, physiological, and anatomical traits of worker honey bees—of pollen hoarding and then genetically map the traits to reveal the underlying genetic architecture, which is an extensive network of interacting genes having effects on many correlated traits.

Mapping the phenotypic and genetic architectures suggested that foraging division of labor in honey bees is derived from networks of genes and hormones involved in regulating reproduction, giving rise to the reproductive-ground-plan hypothesis of Chapter 7. Selection for pollen hoarding has changed the reproductive anatomy of workers, and Chapter 8 shows how the developmental mechanisms that give rise to the different female castes (workers, and queens) were used to generate workers with different-sized ovaries, which in turn affect their responses to environmental stimuli and thus their behavior as foragers and the amount of stored pollen.

Chapter 9 is an attempt to use what we know about stimulus-response relationships, pollen and nectar foraging, and the phenotypic and genetic architectures of pollen hoarding to build a model of the regulatory architecture of pollen hoarding. It is obviously a far more complex task than I can actually manage, but I hope that the process will serve to illuminate what we don't know and help guide future research. Chapter 10 completes the organizational structure of a scientific paper that my mentor Harry Laidlaw taught me: (1) tell them what you are going to tell them (Chapter 1); (2) tell them (Chapters 2–9); and (3) tell them what you told them (Chapter 10).

1.1 Natural History of the Honey Bee

A honey bee colony typically consists of 10,000 to 40,000 worker bees, all females; zero to several hundred males (drones), depending on the time of year; and a single queen, the mother of the colony (Figure 1.2). The nest is usually constructed within a dark cavity and is composed of vertically oriented, parallel combs made of wax secreted by the workers (Figure 1.3, upper panel). Each comb can contain thousands of individual hexagonal cells on each of the vertical surfaces. The individual cells of the combs serve as vessels for the storage of honey (the carbohydrate food source for bees) and pollen (the source of protein) and as individual nurseries for developing eggs, larvae, and pupae. In addition, the combs are the social substrate of the colony. The nest has an organizational structure that is similar to concentric hemispheres, only expressed in vertical planes,

Figure 1.2. A natural swarm of honey bees. Photo by Kathy Keatley Garvey.

where the innermost hemisphere contains the eggs, larvae, and pupae (the brood), the next hemisphere above and to the sides of the brood contains the stored pollen, and the upper and outer regions contain honey that is derived from the nectar of flowers. If you remove a comb that is near, but to the side of, the center of the nest, it will contain three bands covering both sides: the outer band will be honey, the center band pollen,

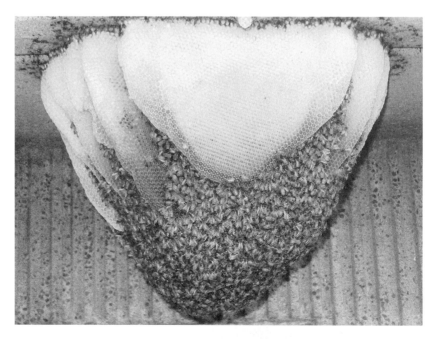

Figure 1.3. Upper panel: A natural honey bee nest with exposed combs. Lower panel: A comb taken from the brood nest of a commercial honey bee colony. The central hemisphere contains capped pupae. Just outside the capped pupae is a band of pollen, and the upper part of the comb contains sealed honey. Photos by Kim Fondrk.

and the lower central part of the comb will contain the brood (Figure 1.3, lower panel). Colonies regulate the amount of pollen, as Jennifer Fewell and Mark Winston showed. They added pollen to colonies and then looked at the effects on pollen foraging and the amount of pollen brought into the hive (intake). Colonies reduced the intake of pollen until they consumed the extra pollen. When pollen was removed from colonies, pollen intake increased until the level of stored pollen was restored.

In addition to the social and nest structures, there is also a structured division of labor. Aristotle pointed this out more than 2,300 years ago (Figure 1.4). He noticed that bees that foraged had less body hair

Figure 1.4. In *Historia animalium,* Aristotle proposed that there was an age-based division of labor where older bees cared for brood inside the nest and younger bees foraged. This was due to his anthropocentric view of puberty. In humans, young people have less pubescence than older people (but then we need to consider "much" older people). However, in bees it is the opposite. Young bees are very pubescent; their thorax is covered by branched hairs. As they age, the hairs break off and are not replaced. Drawing by Sabine Deviche. Used with permission from the Berlin-Brandenburgische Akademie Der Wissenschaften, Page RE, The "spirit of the hive" and how a superorganism evolves, 513–532, 2010.

than those that tended the larvae in the nest. He concluded that there must be a division of labor based on age, with older bees tending the nest and younger bees foraging. He was correct in his assessment of an age-based division of labor, but because of his anthropocentric view of maturation and puberty, he had the age relationship reversed. Newly emerged bees are covered with branched hairs that break off as they age. Foragers are older and often look shiny and nearly hairless. As bees age, they progress through changes in their location in the hive and the behavioral tasks they perform. When they first emerge from their cells as adults, they engage in cleaning cells in the brood nest. When they are about a week old, they feed and care for larvae. This activity is followed by tasks associated with nest construction and maintenance, food processing, receiving nectar from foragers, and guarding the entrance, among others. In about their third or fourth week of life, they initiate foraging. As foragers, they tend to specialize in collecting pollen or nectar; their specialization is demonstrated by a bias in the amount of each they return to the nest. Once they initiate foraging, they seldom perform any within-nest tasks for the duration of their short lives of 5 to 6 weeks.

1.2 Summary Comments

In this chapter, I introduced the concept of the spirit of the hive and the unresolved questions of Maeterlinck about what it is and where it resides. In the next chapter, I will describe the mechanistic processes through which coordinated activities are orchestrated by a "crowd of bees working in a dark hive."

Suggested Reading

Darwin, C. 1998. *The Origin of Species by Means of Natural Selection; or, The Preservation of Favored Races in the Struggle for Life*. New York: Modern Library.

Fewell, J. H., and Winston, M. L. 1992. Colony state and regulation of pollen foraging in the honey bee, *Apis mellifera* L. *Behav. Ecol. Sociobiol.* 30:387–393.

Maeterlinck, M. 1913. *The Life of the Bee*. New York: Dodd, Mead, and Company.

Page, R. E. 2010. The "spirit of the hive" and how a superorganism evolves. In *Berlin-Brandenburgische Akademie der Wissenschaften Jahrbuch 2009*. Berlin: Akademie Verlag, pp. 513–532.

Seeley, T. D. 1995. *The Wisdom of the Hive: The Social Physiology of Honey Bee Colonies*. Cambridge, MA: Harvard University Press.

Winston, M. L. 1987. *The Biology of the Honey Bee*. Cambridge, MA: Harvard University Press.

— 2 —

What Is the Spirit of the Hive?

One cannot observe a hive of honey bees without getting the feeling that they are engaged in highly coordinated and cooperative behavior. As discussed in Chapter 1, both Darwin and Maeterlinck struggled with how this can occur. It seems that there must be some kind of central control, but on careful examination, none can be found. This led Maeterlinck to invoke the "spirit of the hive." But what is it? I will show here that the coordinated behavior long observed and admired emerges from a simple logic of self-organization and requires only that worker honey bees respond to stimuli that they encounter; when they respond, they change the amount of stimulus at that location and thereby affect the local behavior of their nestmates (Figure 2.1).

2.1 Stimulus-Response Basis of Behavior

Stored pollen inhibits foragers from collecting pollen, while young larvae stimulate pollen foraging. Young larvae produce a mixture of chemicals, called brood pheromone, that is secreted onto the surface of their bodies. It is the brood pheromone that stimulates pollen-foraging behavior. Pollen foragers returning from a foraging trip seek out combs with brood and pollen. As the bees walk along the margins of the comb, they come into direct contact with the pheromone and pollen stores. Empty cells indicate that pollen has been consumed by nurse bees (bees that feed and care for the larvae) and fed to developing larvae. Foragers

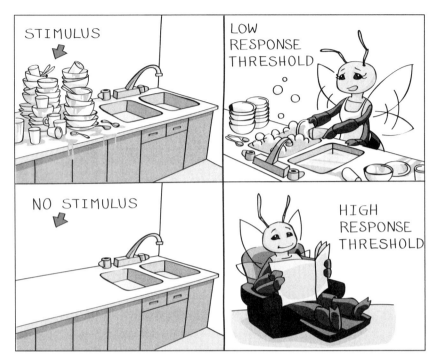

Figure 2.1. Cartoon of the relationship among stimulus, response threshold, and the effect on the stimulus of responding. In this cartoon, a division of labor has emerged. Drawing by Sabine Deviche.

apparently use information obtained from this direct contact to assess colony need.

Let us imagine that pollen foragers have response thresholds to empty cells encountered along the brood/pollen boundary. If a forager encounters more empty cells than some value representative of her response threshold, she will leave the hive and collect another load of pollen. If, however, she encounters fewer empty cells, she does not continue to forage for pollen. Perhaps she switches to nectar or water foraging. This is a very simple view, but it is not unsupported. Tom Seeley in his book *The Wisdom of the Hive: The Social Physiology of Honey Bee Colonies* reported unpublished results of Scott Camazine showing that the number of cells inspected before unloading increased with more stored pollen. In addition, the probability that a pollen forager performed a recruitment

dance decreased with more stored pollen. This shows that pollen foragers are able to make assessments of local pollen stores. Figure 2.2 shows
a cartoon of a returning pollen forager assessing empty cells on a comb.
She has a response threshold of 20 empty cells. If she encounters 20 or
more empty cells, she will unload and make another foraging trip. If

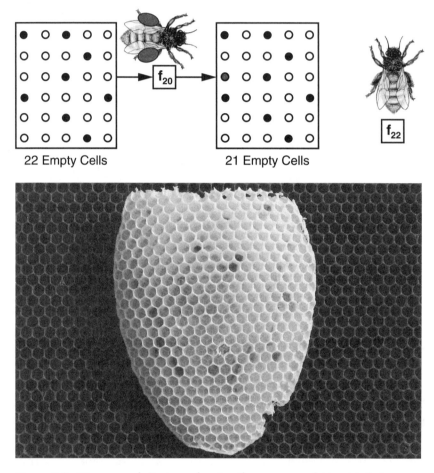

22 Empty Cells   21 Empty Cells

Figure 2.2. Upper panel: Cartoon showing the response relationship between
the stimulus level (empty cells) and behavior, and the correlation between
behavior and the stimulus. Lower panel: Actual comb from colony showing
stored pollen. Used with permission from the Berlin-Brandenburgische
Akademie Der Wissenschaften, Page RE, The "spirit of the hive" and how a
superorganism evolves, 513–532, 2010.

she encounters fewer than 20, she will stop foraging for pollen. The other individual has a threshold of 22 cells. The pollen forager unloads her pollen and then makes another trip. By unloading her pollen, she changes the pollen-stores stimulus from 22 empty cells to 21, which now is below the pollen-foraging response threshold of the other individual. Thus by responding to the stimulus—the number of empty cells—the forager decreases the stimulus by depositing pollen and affects the probability that other individuals will engage in that task. Although this is an oversimplified example, it demonstrates the fundamental basis of self-organized division of labor in the nest.

2.2 The Logic of Division of Labor

The organizational structure of a colony of honey bees is a consequence of a logic that emerges from the behavior of individual bees responding to the stimuli in their environment. The environment of each individual is both instantaneously and experientially unique. The logic of individual behavior is the stimulus-response relationship of an individual with its environment. The logic of a colony is the collection of the stimulus-response relationships of all individuals, the ways in which stimulus information is sampled and shared. We can explore this logic with network models that capture many of the features of real colony behavior.

Models are representations of objects or concepts. They can be scaled-down representations of the real world, like a model of an automobile or airplane, or a map as a model of a physical location. In science, models are most often used to make predictions about some relatively complex system in order to test hypotheses about the model's components, make predictions, or be able to control the parts of the system. In our world today, climate warming is a major issue, and many models have been constructed to better understand the important variables, predict the climatic course of the future, and make recommendations for future courses of action. In agriculture, models are made to estimate the growth of populations of pest insects on crops in order to predict the best time to treat with insecticides and avoid economic damage to the crop. Predictive models are usually constructed with many variables, such as, in the case of crop-damage models, daily average temperature, sunlight, soil moisture, age of the crop, and reproductive and developmental biology of insects. Which of these variables are necessary or sufficient is often unknown and irrelevant to the ability of the model to make reasonable predictions. Alan Gelfand and Crayton Walker called these models "Mulligan stew" models in their book *Ensemble Modeling* after a culinary dish reputed to have been made by unemployed, homeless hoboes in the United States at the turn of the twentieth century. It was a stew concocted from whatever ingredients were available. The individual ingredients were not important, only the taste and nutrition of the end product.

Explanatory models are designed to understand how something works and to seek out those variables that are necessary and sufficient to explain the system. Gelfand and Walker call models of this type "stone-soup" models after a contested story of European origins in which hungry travelers came to a village and asked for food. The villagers were unwilling to share their food, so the travelers set up a cooking pot over a fire in the central square of the village and started boiling a stone. A curious villager asked them what they were doing, and they explained that they were making stone soup but were missing a few ingredients needed to improve the flavor. The villager shared some small item of food. Another curious villager came by, asked the same question, and shared some other small item of food. Eventually, the travelers had a soup that was tasty and nutritious. Stone-soup models start with the minimum parameters and add more only as needed to explain the system studied.

Explanatory models can be built and then tested by comparing a specific outcome with the model's predictions based on hypothesized relationships among a set of variables. The models are modified and sequentially retested until a good fit is achieved, as in the case of optimality models in evolutionary biology that attempt to predict an optimal phenotype (the observable traits of an individual, or a colony in the case of social insects) for a given environment given a specific set of parameters and constraints. Or groups of models can be constructed where the parameters of a model are systematically varied throughout plausible ranges and general behavioral features of the system are explored, an ensemble-modeling approach.

Directed graphs have been used to study complex systems that can be expressed as networks. It is easy to imagine a honey bee colony as a network of individuals, each making independent decisions based on local information processed through a brain of 900,000 neurons that has been programmed by development and experience. No single individual has complete information about the current state or history of the colony, but collectively the 30,000 individuals with 900,000 neurons each (a total of about 27 billion neurons, less than half the number in a human brain) are capable of storing an enormous amount of information about the hive and foraging environments. Individual workers share a limited amount of information with their nestmates. Usually this is shared indirectly through their behavioral activities, but some information is shared directly through an intricate communication system involving chemicals (such as pheromones), shared food, and their recruitment behavior—called the dance language of the bee.

Sandra Mitchell and I used a directed-graph-network, ensemble-modeling approach to explore the origins of division of labor. In early 1989, I was invited to be the token biologist in the philosophy of biology symposium at the biennial meeting of the Philosophical Society of America, held in 1990. The symposium featured philosophers and historians who were familiar with the work of Stuart Kauffman (Kauffman was trained in philosophy as an undergraduate) and who could especially comment on his new book *The Origins of Order*. Kauffman used ensemble modeling and directed graphs to model gene regulation in the 1970s. I was asked to read a draft of his manuscript and see whether I found his book to be of any value to me, a biologist, in looking at the level of social organization. I agreed and immediately called my good friend Sandra Mitchell, a philosopher of biology and an acquaintance of Kauffman's, and asked her to help me. I needed help because I had no idea what to expect from philosophers. She agreed, and we went to work building a model.

We assumed that a group of nonsocial but tolerant insects lived together in a nest. This is very likely because there are many nonsocial, nesting insects that lay eggs in batches, or in close proximity, within a common nest. The adult offspring that emerge in the nest begin their lives together at the same location. A directed-graph network of a group of protosocial insects such as these might look like Figure 2.3. The dots are called *elements* and may represent individual insects. Elements can be nodes for information flow. The arrows are informational channels. The information begins at the node at the tail of the arrow and ends at the head of the arrow at another node. In the simplest network, the arrows transmit information about whether the originating element is "on" or "off." This can be expressed in Boolean logic as 0 or 1. In our protosocial insect case, this could be an individual performing or not performing a given behavior. For instance, an individual foraging for pollen would be a 1, while another individual not foraging for pollen would be 0, although she could be a 1 for some other task, such as foraging

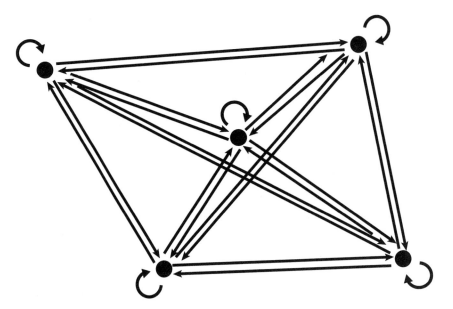

Figure 2.3. Network graph in which there are five nodes, each connected to all other nodes and to itself. Information travels from one node to the next through the directed arrow in the direction of the arrowhead. In this case, $K = N = 5$.

for nectar. Graphs consist of N elements forming nodes with a connectedness K, the average number of arrowheads terminating at each elemental node. Connections can be regular, random, or however you want to make them. Figure 2.3 has $N = 5$ and $K = 5$ (with symmetry of every element connecting to every other element, including itself). Each element, in turn, must have a "decision" function to determine whether it is on or off on the basis of the inputs it gets from the other elements to which it is connected. In set notation from high-school algebra, the set of all functions assigned to all elements is $\{F\}$. One kind of function is a threshold, for example, "If three or more of the elements that are connected to me are on, then I will be off." Therefore, I have a threshold of 3. Threshold functions are a special subset of Boolean functions (Table 2.1).

Our stone-soup ensemble model was developed for computer simulation. We randomly assigned a threshold decision function to each of the elements. Thresholds were drawn from a predetermined distribution of the subset of Boolean threshold functions. For example, each element was randomly assigned a number between 0 and N. If the number of "on" inputs to the node (reflecting the states of the nodes to which the inputs are connected) is equal to or greater than the assigned number, then the element turns off. If the number is less than the as-

Table 2.1 Boolean functions for $K = 2$

Boolean functions

X_1X_2	1	2	3	4	5	6	7	8	9	10	11	12	13	14	15	16
00	0	1	0	0	0	1	1	1	0	0	0	1	1	1	0	1
01	0	0	1	0	0	1	0	0	1	1	0	1	0	1	1	1
10	0	0	0	1	0	0	1	0	1	0	1	1	1	0	1	1
11	0	0	0	0	1	0	0	1	0	1	1	0	1	1	1	1

Note: Boolean logic is based on sets of functions that can exist in only two states, 0 or 1. In this example, "off" is assigned to the case of 0 and "on" to the case of 1. There are only two inputs to each node, X_1 and X_2. There are four possible input states for nodes with two Boolean inputs, as shown in the column on the left: 00, 01, 10, and 11. There are 16 possible functions. For example, function 1 is "off" regardless of inputs, while function 3 returns a 1, "on," only when $X_1 = 0$ and $X_2 = 1$. Function 2 is a threshold function that returns a 0 if either of the two inputs is 1.

signed number, it turns on. We assume that there is some level of stimulus in the nest that needs to be regulated or eliminated, for example, heat. If the temperature is too hot, the insects might fan their wings to cool the nest, a task regularly performed by honey bees. The model assumes that some level S of stimulus is experienced by the model insects, represented by the elements. When an individual is "on," it decreases the stimulus (cools the nest) by some constant amount, $1/S$. The number of elements that are on is called the *density (D)* of the network. The current stimulus level (S_t) experienced by the elements (our network model insects) is equal to the total initial stimulus level (S_0) minus the number of individuals that are currently on, so $S_t = S_0 - D$. This is equivalent to a directed-graph network with all nodes connected to all other nodes, as in Figure 2.3. On the basis of a predetermined method of sampling the elements in the network, sampled elements first turn off (stop performing some task), which then increases the stimulus level, S_t, by one unit. The element then checks the current stimulus level, S_t (an indirect way of determining the states of all other elements in the network represented by D) and makes a decision to be on or off on the basis of its threshold function. If it turns on, S_t is decreased by one unit.

This is the stone-soup model: binary behavior, Boolean logic, and random assignment of decision functions. We looked at ensembles of these models. With the ensemble-modeling approach, we generated many models by varying N, K, $\{F\}$, and S_0 over their plausible ranges. We set $N = 100$ or $1,000$ and $K = N$ and varied the threshold set with respect to its distribution. We looked at general (global) properties of the system rather than specific outcomes, such as the ability of different combinations to regulate the stimulus level at some value (equilibrium), how long it took the network system to reach the regulated equilibrium, and the activity (number of elements that turned on or off) of the network at equilibrium. We also incrementally increased S to test the ability of the network to respond and regulate the increasing stimulus level (equivalent to turning up the heat in our example) and varied the way in which the elements sampled the environment: we assigned elements numbers, randomly selected elements one at a time, sampled them in sequential order, or sampled all simultaneously. The global behavior of the model network was similar to the global behavior of insect societies (Figure 2.4), but this is still just a computer model. How does it map onto a real colony of honey bees?

In Section 2.1, I presented a model of how a pollen-foraging honey bee might be able to get information about the pollen need of a colony by walking along the edge of a comb where the uncapped larvae border the stored pollen. I suggested that thresholds of response to empty cells or cells filled with pollen that a forager encountered could determine whether she would return for another load of pollen. That case is similar to the model presented here. Foragers get indirect information about the behavioral states of others by sampling the database of the pollen foragers, the stored pollen. Stored pollen is a stimulus that inhibits pollen foraging. Consumption of pollen by nurse bees decreases the inhibitory stimulus. The result of the network of foragers is the maintenance of pollen stores at or above some level determined by the collective response thresholds of the foragers. Foragers, however, cannot and do not get complete information about the quantities of stored pollen, so $K < N$. Instead, they take a subsample as they walk the comb and then decide whether to perform a recruitment dance and return to the field.

2.3 Case Studies

The simple result of these models is that division of labor is an emergent property of group living. It is inescapable. Behavior is based on responses to stimuli. Individuals differ in their thresholds of response and the probability that they will encounter the stimulus, and as a consequence of their response, they alter the stimulus and thereby affect the response probabilities of others in the group. Is there any evidence for this view other than what we found with our computer simulations? To test this hypothesis, one must force social conditions on otherwise nonsocial individuals or generate conditions in a social species that are not normally found and observe the responses. I present two case studies here.

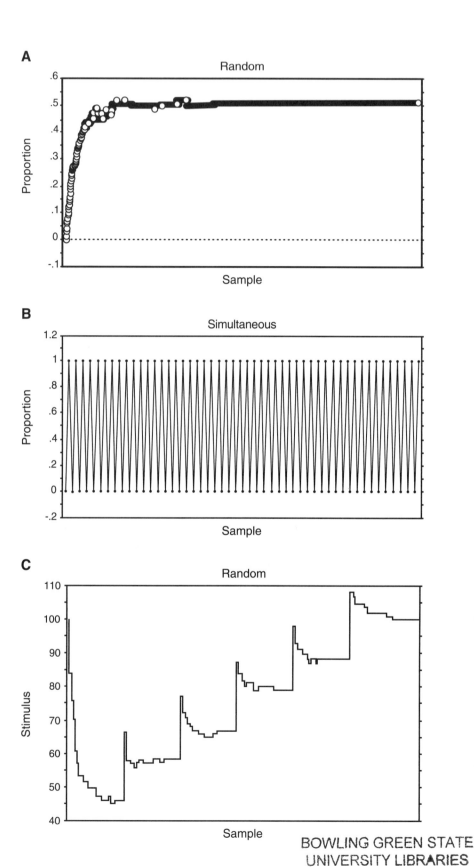

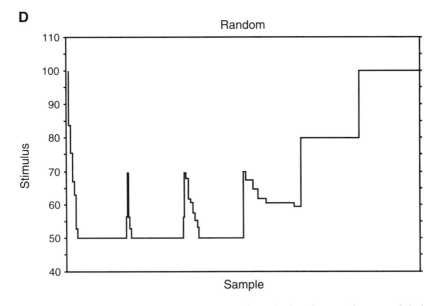

Figure 2.4. Results of computer simulations described in the text showing global system behavior. For each sampling event of the simulation, the proportion of individual elements that are "on" is shown (Panels A and B) or the residual stimulus is shown (Panels C and D). Panel A: The distribution of thresholds among elements is random between 1 and 100, with an average of 50.5. Initially, the stimulus level was constant at 50 and was decremented 1 unit when an element turned on. This model demonstrates properties of homeostasis resulting from negative feedback regulating the stimulus level at a point determined by the average of the threshold distribution used, such as occurs, perhaps, with honey bee thermal regulation of the nest. Panel B: The distribution of thresholds among elements is random between 1 and 100, with an average of 50.5. Initially, the stimulus level was constant at 50 and was decremented 1 unit when an element turned on. The stimulus level was sampled by all elements simultaneously, resulting in mass-action responses of all elements with thresholds below the initial stimulus level turning on and off simultaneously. This action is similar to alarm responses in honey bee colonies. Panel C: The distribution of thresholds among elements is random between 1 and 100, with an average of 50.5. Initially, the stimulus level was constant at 50 and was decremented 1 unit when an element turned on. The stimulus level was increased 20 units after every 500 samples of the network. The network displayed system plasticity and resilience in response to increases and decreases in stimuli such as we observe in undertaking behavior (see Section 3.1 for discussion of plasticity and resilience). Panel D: The distribution of thresholds among elements is fixed at the average value of the simulations, 50.5. All elements operate with same threshold. Initially, the stimulus level was constant at 50 and was decremented 1 unit when an element turned on. The stimulus level was increased by 20 units after every 500 samples of the network. This network demonstrated a pronounced division of labor among the elements, with some frozen on while others were frozen off. This behavior might be observed when a finite number of worker honey bees respond to an increasingly large stimulus. From Page and Mitchell 1998, *Apidologie* 29(1–2): 171–190, Fig. 3, http://dx.doi.org/10.1051/apido:19980110.

2.3.1 *Case Study of* Ceratina flavipes

Japanese researchers Shoichi Sakagami and Yasuo Maeta worked with a small carpenter bee called *Ceratina flavipes*. Females excavate the pithy centers of stems and then provision cells with loaves of pollen mixed with nectar, beginning at the point farthest from the entrance. After completing a pollen loaf, the female lays an egg on it, seals the cell, and begins foraging to provision the next cell. *Ceratina flavipes* females are solitary. Females guard the entrance of the nest from parasites and predators by plugging the entrance hole with their bodies. Sakagami and Maeta forced pairs of females to nest together in cages by restricting the number of twigs they could excavate. After making attempts with 178 females, the researchers were successful in getting only five pairs to nest together. In each of these five cases, one female became the primary egg layer and guarded the entrance, while the other foraged. A division of labor emerged between what are normally two solitary individuals.

2.3.2 *Case Studies of* Pogonomyrmex spp.

Jennifer Fewell did a similar experiment with ant queens from the species *Pogonomyrmex barbatus*. *P. barbatus* queens begin excavating a nest in the sandy soils of southern Arizona soon after they make their dispersal flights from the natal nest and mate. They shed their wings and then begin digging. They initiate nests on their own. Studies of one population in Arizona have been ongoing for more than 30 years without a single observation of two queens inhabiting the same nest. Therefore, they are strictly solitary in their nest-founding behavior. Fewell placed 63 pairs of individually marked, new queens in glass vials containing sand and observed their digging behavior (Figure 2.5). Nest excavation took place in all vials. In most cases, there was a significant difference in the amount of time the two queens forced to cohabit spent digging. In other words, a digging division of labor emerged.

Fewell pretested 18 solitary *P. barbatus* queens to determine their propensity to dig. She then combined them in pairs, observed their

Figure 2.5. A young mated *Pogonomyrmex barbatus* queen, shortly after shedding her wings, begins to excavate her founding nest chamber. Drawing by John Dawson/National Geographic Stock.

rates of excavation, and found that in eight of nine pairs the queen that did more digging when nesting alone also did more digging when paired. In addition, the queen of the pairs that dug less when solitary showed a significant reduction in digging behavior when she was paired with a queen with a higher propensity to dig. This looks like an example of spontaneous division of labor with amplification of the differences in behavior between the queens due to their responses to the actions of each other (Figure 2.6).

Fewell set up parallel experiments using queens from a different species, *Pogonomyrmex californicus*. In some populations, *P. californicus* queens form cooperative groups of up to 30 queens and cofound new nests. With an evolutionary history of cofounding, one might expect differences with respect to division of labor. However, division of labor for digging (digging asymmetry) was no stronger in *P. californicus* than that observed for *P. barbatus*. This suggests that an evolutionary history of nest sharing is not necessary to get a demonstrable division of

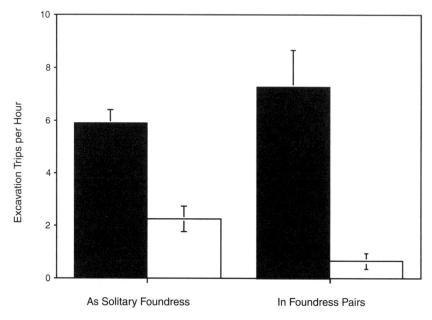

Figure 2.6. *Pogonomyrmex barbatus* queens were tested as solitary nesters for their activity rates when they engaged in excavation behavior and were rated as high-frequency (black bars) or low-frequency (open bars) excavators. Low-frequency excavators decreased their excavating when they were forced to found nests with high-frequency excavators while high-frequency excavators increased their excavating. Thus their behavior influenced each other and amplified the difference. From Fewell and Page 1999, "The emergence of labor in forced associations of normally solitary ant queens," *Evolutionary Ecology Research* 1:537–548, Fig. 3.

labor. As shown by the network model, some social organization emerges just from the structure and functional elements of the system, in this case the proximity of individuals, the differences in response thresholds, and the effect of the response behavior on the stimulus. More complex social systems evolve by adjusting N, K, and $\{F\}$.

2.4 Adaptive Fine Tuning of Division of Labor

A division of labor among cohabiting individuals is an inescapable property of group living and the stimulus-response basis of behavior.

But how does social behavior evolve to be more complex? In Chapter 5, I dig deeper into the complex social behavior of food foraging and storing and show how we successfully changed the social organization of two populations of bees by artificial selection. How does selection affect the social system? Selection must change the dynamic systems parameters of N, K, and $\{F\}$.

Selection on N could increase the total number of individuals in a colony or the numbers that are competent to perform a task. For example, queen fecundity could increase. For a given, fixed life span of workers, a higher egg-laying rate by the queen would lead to a larger number of workers in the colony. Or the life span of workers could be extended. Then for a given egg-laying rate of the queen, a colony would have more workers when it reached the equilibrium point of births equal to deaths.

Selection on K affects the connectedness of individuals either directly or indirectly through their effects on other stimuli. Selection on K could affect the sensitivity of sensory sensilla (special anatomical structures that perceive stimuli) to specific stimuli or affect the motor patterns of workers or their location in the nest, making more bees likely to come into contact with specific task-releasing stimuli.

Selection on $\{F\}$ affects the distribution of response thresholds of workers in colonies. This could include the average thresholds of individuals in a colony or the variance in thresholds. The distribution of thresholds affects the dynamics of model network colonies, even when the average thresholds remain the same. Honey bee colonies with more genetic diversity are thermally more stable than those of lower diversity. Selection favoring a general increase in genetic variation, for example, through multiple mating of queens (polyandry), increases the variance in response thresholds when there is heritable genetic variation. Also, stimulus-response thresholds can be grouped and change with age or experience, perhaps resulting in task sets for individuals that can change with time. Network models where elements make decisions to be on or off to two tasks at a time show more dynamic stability than single-stimulus-threshold models.

2.5 From Stone Soup to Mulligan Stew

The directed-graph-network model I have presented is obviously incomplete, although I believe that it captures the essential elements of emergent complex social behavior. In the sections that follow I discuss the features of the model that can be altered to better reflect reality but also result in the model being perhaps less general and understandable.

2.5.1 Decision Functions Are Not Fixed

The model assumes that the threshold functions of individual elements are fixed. However, learning experience affects response thresholds and the probability that an individual engages again in a given task. Another effect is to generate greater variance in the distribution of thresholds in a colony {F} and affect the colony's behavioral dynamics. One example is the response of a colony to odors it has learned are associated with specific food sources. Odors that initially have no effect on recruiting experienced foragers recruit large numbers of foragers when they are presented inside the nest after foragers have experience with them.

2.5.2 Response Thresholds Change with Nutrition

Jochen Erber and I performed a set of experiments that demonstrated differences in response thresholds to sucrose between pollen and nectar foragers. I was on sabbatical leave at the Technical University of Berlin. I had been invited to give a lecture at an international conference and wanted to discuss some elements of the response-threshold models I had been working on for 10 years. I needed a demonstration of a response threshold for my talk. Until this time, I had treated response thresholds as black-box phenomena that must be true. How else can one explain behavior? Erber suggested that we collect some foragers at the entrance of a colony in the Technical University's apiary and test them using the proboscis extension response (PER) test (Figure 2.7). He would take pictures. We subjected bees to increasing concentrations of sucrose solutions. When we went to collect the foragers, we decided to do an experiment and compare pollen and nectar foragers (just for fun). We did not think that they would actually be different. We were surprised to find that the pollen foragers responded significantly more

Figure 2.7. Proboscis extension response test of sucrose sensitivity. A sucrose-solution concentration series was prepared, usually with concentrations of water only and 0.1, 0.3, 1, 3, 10, and 30 percent sucrose. Bees were placed in confining tubes with their heads exposed. Droplets of the different solutions were touched in increasing concentrations to each antenna to stimulate the extension of the proboscis. The response threshold was estimated by the lowest sugar concentration that resulted in the extension of the proboscis. We often also refer to "the responsiveness of the bee." Responsiveness is just another way to talk about response thresholds because bees do not respond to sugar solutions with concentrations below their threshold and respond to all or most solutions above it. A count of the total number of times they respond is a measure of their responsiveness and indirectly their response threshold. We call the total count the gustatory response score (GRS). A high GRS corresponds to a low response threshold, and vice versa. Photo by Jochen Erber. Reprinted from *Current Topics in Developmental Biology* 74, Page et al., "The Development and Evolution of Division of Labor and Foraging Specialization in a Social Insect (*Apis mellifera* L.)," 253–286, Fig. 2 (2006), with permission from Elsevier.

to water and to lower concentrations of sugar than the nectar foragers. They had lower response thresholds. We thought that this might be because they were more motivated by hunger. Nectar foragers are full of sugar solution and satiated, while pollen foragers are under nutritional stress from flying to collect pollen. We fed all bees to satiation with 30 percent sucrose solution, let them rest for one hour, and then retested them. Responses to water were no longer different and responses to sugar solutions were greatly reduced in the satiated bees, but the differences between pollen and nectar foragers persisted. In fact, the relative differences between them increased. Response thresholds were not fixed; they had been modulated by feeding.

Tanya Pankiw and Keith Waddington kept young bees in cages in an incubator and fed different concentrations of sugar solution to bees in the different cages. Bees fed higher concentrations of solution had higher response thresholds (were less sensitive to sugar) than those fed lower concentrations. Bees foraging at feeders of different concentrations modulated their response thresholds according to the concentration of the solutions they collected. As with the bees in cages, those that foraged on more concentrated solutions were less responsive than those that foraged on solutions with lower concentrations. The bees were collected before they fed at the feeders; this showed that the modulation of response thresholds was based on their foraging experience, not on their level of satiation at the feeder. The filling of the crop (Figure 2.8) also had an effect on responsiveness, but it was ephemeral. Bees collecting water arrived at the water feeder with a low response threshold, but their response thresholds increased as they filled. Because water has no nutritive value, their response was, again, a consequence not of nutritional satiation but probably of the mechanical filling of the crop.

2.5.3 Response Thresholds Change with Experience

Nectar foragers pass their nectar loads to receiver bees when they return from a foraging trip. (Receiver bees are preforaging-aged bees that re-

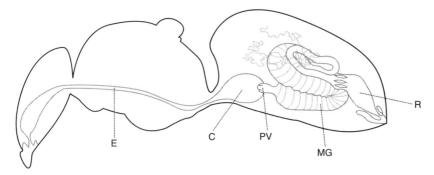

Figure 2.8. Diagram of the digestive tract of a honey bee. It is composed of three chambers, the honey stomach, or crop (C), the midgut (MG), and the rectum (R). The proventricular valve (PV) sits between the crop and the midgut. The esophagus (E) links the mouth and the crop. Drawing by Adam Tofilski, honeybee.drawwing.org.

main near the entrance of the colony or the dance floor and collect the nectar loads from the returning foragers. They then either deposit the nectar in cells or distribute it to other bees.) They also pass some nectar to bees that attend their dances, and these bees in turn share it with others, and so on. The incoming nectar is distributed and modulates the response thresholds of bees throughout the nest. Mindy Nelson and Tanya Pankiw set colonies in flight cages and provided the cages with sucrose-solution feeders. They changed the sucrose solutions in the different cages and then collected nonforaging young bees and tested them for their response thresholds to sucrose solutions. The young bees' responses were modulated by the concentration of the sugar solution collected by the foragers. When the concentration was high, their response thresholds were high. When the concentration was low, their response thresholds were low. These results demonstrate that a colony of bees "shares a common stomach" and that the bees in the hive have the capability to track the quality and perhaps the quantity of the nectar being collected in the field. This is also another demonstration of how all the bees in the colony are connected indirectly in a network and are able to get information about the global state of the network.

Receiver bees locate themselves on the "dance floor" of the hive, situ-ated on brood combs near the entrance, and unload the nectar from the foragers when they return from a foraging trip. The receiver bee stands head-to-head with the forager and extends her proboscis. The forager opens her mandibles and expresses a droplet of nectar that the receiver bee sucks into her crop. Sucrose-response thresholds of receiver bees are modulated by nectar received from foragers. Receiver bees that un-load nectar of lower sugar concentration are more responsive to sugar. The sucrose-response thresholds of other nonforaging bees are differ-entially affected by the flow of sugar solution in the hive. However, nurse bees, which are younger on average than receiver bees, are less affected by the incoming nectar. The presence of an odor in sugar solution ampli-fies the modulation of sucrose-response thresholds in 14-day-old bees, the age they are likely to be receiver bees, but not in 7-day-old bees, which are likely to be nurses.

Bees that perform different tasks have different response thresholds to sucrose solutions. José Pacheco and Mike Breed collected bees that were engaging in different tasks: foraging for pollen, foraging for nec-tar, foraging for water, foraging for pollen and nectar, fanning (a behav-ioral act that can serve in thermoregulating the nest), guarding the entrance from intruders, and undertaking (the removal of dead bees from the nest). They found consistent differences across colonies in su-crose responses of bees engaged in the different tasks.

2.5.4 Response Thresholds Vary with Age
and Exposure to Pheromones

Responses to sucrose change with age. Tanya Pankiw tested bees for their responses to sucrose solutions before they initiated foraging. Bees were of three age classes: (1) less than 48 hours old, (2) more than 48 hours but less than 1 week old, and (3) more than 1 week but less than 2 weeks old. She found that responsiveness to sucrose increased with age.

Pheromones are chemicals commonly used by insects for communi-cation. Honey bees produce many different pheromones that aid in the cohesion of the social unit and regulate the response of colony members to changing conditions. For example, isopentyl acetate is a compound produced by the sting gland of worker honey bees that releases stinging behavior. It is known as the alarm pheromone. Queen honey bees produce a blend of compounds called queen pheromone. The main component is 9-oxo-2-decenoic acid, which is produced in the mandibular glands of the head. This pheromone blend has multiple functions, including suppressing egg production and oviposition (egg laying) of workers, inhibiting the production of new queens by colonies, and serving as an attractant for males (drones) when queens make their mating flights. Larvae produce a blend of compounds that they secrete to the body surface and that stimulate nurse bees to feed them. The so-called brood pheromone also releases pollen-foraging behavior in foraging-aged workers.

Exposure to queen mandibular pheromone and brood pheromone reduces responsiveness of bees to sucrose solutions. Social pheromones such as these modulate behavior and response thresholds.

2.5.5 Hormones, Biogenic Amines, and Stress Modulate Response Thresholds

Juvenile hormone (JH) is an important growth regulator in insects and also is involved in changes in reproductive physiology and behavior. Juvenile hormone has long been known to affect age-related changes in behavior. In honey bees, it is part of the regulatory hormonal network that results in the transition of worker honey bees from performing tasks in the nest to foraging. Topical treatment of newly emerged bees with methoprene, a commercially available synthetic form of juvenile hormone, results in a significant increase in sucrose sensitivity. Octopamine is a compound produced in the brain that modulates sensitivity to sugar. When bees are fed octopamine in sucrose solutions, they become more sensitive to lower concentrations of sugar solutions—it lowers their response threshold to sugar.

Stress also reduces the response threshold to sucrose. Tanya Pankiw placed newly emerged bees in small tubes to test their sucrose responsiveness. Some bees were narcotized with carbon dioxide or cold temperature before they were placed in the tubes, while others served as controls and were placed without narcosis. Bees were tested 30 and 60 minutes later. Bees that were narcotized were less responsive to sucrose than the controls at 30 minutes, but the differences disappeared by 60 minutes. Narcotized bees were shielded from the stress of handling. Stressed bees became more responsive, probably because of elevated levels of octopamine and juvenile hormone. Brain levels of octopamine increase dramatically and rapidly with stress even when the stress is simply the holding of a leg with forceps. Juvenile hormone titers also increase with stress, are known to be affected by octopamine, and affect octopamine levels.

2.5.6 Links of the Sucrose-Response Threshold to Other Sensory Modalities

Sucrose responsiveness correlates with other sensory modalities. The response to antennal stimulation with sucrose correlates with sucrose stimulation of the mouthparts, but response thresholds to sugar presented at the antenna are 10 times higher than at the proboscis. Responses to sucrose based on antennal stimulation also correlate with responses to citral (olfactory), pollen (gustatory), and light (visual).

The relationship between sucrose sensitivity and response to light has been best studied. Jochen Erber developed a light assay where bees were placed in a dark, circular chamber. They remained for several minutes before testing in order to adapt to darkness. The chamber had green-light-emitting diodes in a circle at the base that emitted light at different intensities (Figure 2.9). Lights of equal intensity opposed each other on the circle. Lights were turned on individually, in increasing intensity; each pair of lights of equal intensity doubled the intensity of the previous pair. Activities of bees were recorded by using an infrared light source and a video camera. One light of a given intensity was turned on, and the time for the bee to walk to that light was recorded. When the bee reached the light, it was turned off, the opposing light of the same intensity was turned on, and the bee was timed and recorded

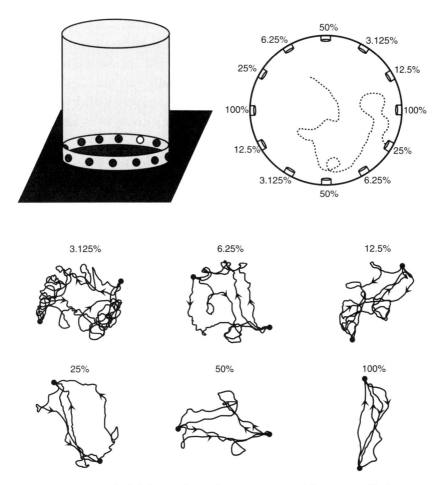

Figure 2.9. Top: The left figure shows the test arena used for assays of light sensitivity based on the design of Jochen Erber. Bees were placed under the light bucket and were exposed to light of increasing intensities. Lights of equal intensities were on opposite sides of the chamber. Relative light intensities increased serially from 3.125 percent to 100 percent. The right figure shows the path of a tested bee. A light was turned on, and the bee was given time to orient to the light and approach it. After the bee reached the light, it was turned off, and the light of the same intensity on the opposite side was turned on. The bee would then traverse the arena to the light on the other side until it made four complete trips. Then the next light intensity was tested. The bottom part shows how the length of the path decreased with greater light intensities. Reprinted from *Behavioural Brain Research* 205(1), Tsuruda JM, Page RE. "The effects of foraging role and genotype on light and sucrose responsiveness in honey bees (*Apis mellifera* L.)," 132–137, Fig. 1 (2009), with permission from Elsevier.

until it reached the light at the opposite side. After a set number of back-and-forth trips, paired lights of the next-higher intensity were used. If bees did not respond to a given light intensity after a set amount of time, the next-higher light intensity was used. Erber and his colleagues found that nectar and water foragers with higher responsiveness (lower response thresholds) to sucrose were also more sensitive to light. Jennifer Tsuruda tested young, nonforaging bees for their sucrose-response thresholds and then used this apparatus to test them for their responses to light. Foragers that responded to lower concentrations of sucrose solution also responded to lower light intensities. She found a significant correlation between sucrose and light responses.

2.5.7 Responses Are Analog

Our stone-soup model assumes a Boolean-binary logic (Section 2.2). Decision functions deliver an on or off result along directed arrows to connected nodes. The logic of the "spirit of the hive" is probably analog (continuous) with suprathreshold responses. In other words, responses vary in intensity with increases in stimulus levels above thresholds.

2.5.7.1 Proboscis Extension   Suprathreshold responses are shown at the most elementary level by looking at the proboscis extension response (Figure 2.7). Bees whose antennae are stimulated with sucrose solution frequently only partially extend their proboscis. For the next-higher concentration they may extend it more slowly, then faster for the concentration that follows. Stephan Haupt, working in the laboratory of Jochen Erber, recorded electrical impulses from a single muscle used by the bee when she extends her proboscis. The number of electrical impulses per second increased linearly with the concentration of sucrose-solution stimulation of the antenna. Merideth Humphries showed that the walking speed of newly emerged honey bees correlates with sucrose responsiveness. Bees that are more sensitive to sugar walk faster. This suggests a more general relationship between the sensory and motor systems of bees.

2.5.7.2 Associative Learning The relationship between sucrose responsiveness and learning is also continuous, not binary. Associative learning is the response of an individual to the pairing of a stimulus with a reward (Figure 2.10). Performance on associative-learning assays is conditional on the sucrose-response thresholds of the bees. Honey bees can be conditioned to tactile stimuli or to odors. To study tactile learning, bees are pretested for their responsiveness to sucrose solutions and are given a gustatory response score (GRS; the total number of times they respond while being presented with the series of sucrose solutions—see Figure 2.7). Each bee then receives an acquisition score, which is the number of unconditioned stimulus (US)–conditioned stimulus (CS) pairings that are necessary before the bee presents a conditioned response. Bees that are more responsive to lower concentrations of sucrose and water take fewer trials to acquire a response to the CS—they have faster acquisition. Thus the perceived value of the reward, not the actual value, determines the strength of learning. The difference between the concentration of sucrose offered and the response threshold of the bee represents the suprathreshold reward value of a sucrose stimulus. A larger differential results in better and more rapid acquisition and better memory retention—a continuous rather than binary function. This is true for both tactile and olfactory learning.

The word *motivation* is often used in behavioral studies, especially in studies of associative learning. A bee that is replete with food is not motivated to learn. A hungry bee is very motivated. Bees that are replete are less responsive to sugar than those that are hungry, and their performance on associative-learning tests reflects this. Their response thresholds are modulated by their nutritional status, and the difference between their responsiveness to sugar and the value of sugar reward affects their learning. Jochen Erber and I defined the difference between response threshold and the reward value as motivation, making it a relative rather than an absolute concept.

2.5.7.3 Nectar and Pollen Load Sizes   Karl von Frisch won the Nobel
Prize in Physiology or Medicine in 1973 for his work on the dance language of the honey bee (Figure 2.11). He did many wonderfully simple
but elegant studies of foraging behavior, looking at the economics of foraging, as well as the communication system of honey bees, and showed
that bees that are offered higher concentrations of sucrose solutions collect larger loads. The rate at which a bee imbibes sugar solution is also
correlated with sugar concentration. If the solution is more concentrated,
they collect it faster at a feeder, and the food is transferred faster to a receiver bee back at the nest when the forager returns. Foragers that are
more responsive to lower concentrations of sucrose will forage at artificial
feeders with a more dilute sugar solution than will less responsive bees.

Pollen and nectar loads are not independent. The nectar load size of
foragers increases with the concentration of the nectar collected. The size
of pollen loads decreases with an increase in nectar load. We collect bees
at the hive entrance as they return from foraging, take them into the

---

Figure 2.10. Tactile learning. A honey bee shows conditioned proboscis
extension response during tactile antennal conditioning. Associative-learning
assays for honey bees are based on standard classical conditioning models where a
stimulus one wants them to learn is paired with a reward. Conditioning is based
on the forward pairing of the conditioned stimulus (CS) with a reward. In tactile
learning, harnessed bees have their vision occluded by black paint and are
presented with a small copper plate with etched grooves, the CS. Their antennae
are naturally active, scanning their surroundings. When the antennae touch and
move over the plate, the bees get information about the characteristics of the plate
from hairlike mechanoreceptors located on the tips of the antennae. After the bee
scans for a few moments, a droplet of sucrose solution (of sufficient concentration
to exceed its response threshold) is touched to the antenna, an unconditioned
stimulus (US), eliciting a proboscis extension response. With the proboscis
extended, a droplet of sucrose is touched briefly to the tip of the proboscis, the
reward, and the bee is allowed to imbibe a small quantity. The brass plate is then
withdrawn. The sequence thus is CS-US-reward. This sequence is repeated for a
number of trials. Usually after a single or a few trials, the bee will respond by
extending the proboscis when it is presented only with the target; the US is no
longer needed. Photos by Jochen Erber. Reprinted from Page and Amdam 2007,
"The making of a social insect: developmental architectures of social design,"
Bioessays 29:334–343, Fig. 2, with permission from John Wiley and Sons, UK.

Figure 2.11. Cartoon of the honey bee waggle dance. The dancing bee in the middle engages in a figure-eight dance that the other potential recruit bees follow. The dancer is currently in the straight-run portion of the dance. She will move to the bottom of the figure eight, turn right or left alternately, go back to the top, and make another waggle run through the center. Drawing by Jacob Sahertian.

laboratory, anesthetize them with carbon dioxide, and squeeze the contents of their crops into a capillary tube. The capillary tube is weighed to determine the size of the load, and the sugar concentration of the nectar is determined with a refractometer. We also remove their pollen loads and weigh them. Many studies conducted in my laboratory over the past 20 years have shown a significant positive correlation between nectar load size and nectar concentration and a negative correlation between nectar load and pollen load (Figure 2.12), demonstrating another suprathreshold response.

2.5.7.4 Recruitment Dances   Recruitment dances are performed in response to the quality of nectar collected in the field and the rate at

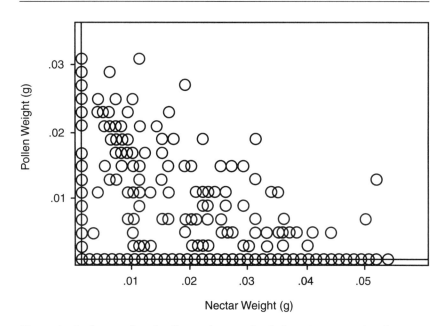

Figure 2.12. Scatterplot of pollen and nectar loads from 369 returning foragers.

which the receiver bees unload the foragers. The probability of performing a dance and the duration of the dances bees perform correlate
with the concentration of the sugar solution they collect. Tom Seeley
has shown how the simple processes of individual assessment of food
quality and unloading time provide information about the nutritional
status of a colony and a mechanism for differential recruitment of bees
to forage for the more profitable resources. The probability, vigor, and
duration of waggle dances correlate with the concentration of sugar
solution collected. Seeley trained individual foragers to a feeder where
he varied the concentration of sugar solution offered. He then observed and recorded the number of waggle runs they produced with
their recruitment dances. He found a continuous linear relationship
between sugar concentration and waggle runs, a suprathreshold, analog response. He also found that each bee he tested had a different relationship between concentration and waggle runs; this suggests that
the sucrose-response relationships were different for each of the seven
bees of the study.

Keith Waddington and colleagues studied the round dance of bees that is performed when bees forage close to the nest. They found that the number of reversals of the dance per minute and the duration of the dance correlated with the calories gained by foragers attending artificial feeders. Calories gained were directly related to sugar concentration delivered at the artificial flower. Properties of the sound produced by round dances also changed with sugar concentration.

Waddington and Mindy Nelson presented high- and low-quality pollen at feeders within a cage. (Pollen can be collected from colonies in large quantities and then fed back during periods of pollen dearth. Returning foragers are forced to pass through a wire grid that is slightly bigger than their bodies. The grid knocks off the pollen loads, which then fall into a collection tray. The pollen pellets are harvested and usually frozen fresh or dried for storage.) Low-quality pollen was freshly ground (in a coffee grinder) pollen that was adulterated with alpha cellulose, an inert solid substance that bees can collect. The high-quality pollen was unadulterated. Bees were more likely to perform round dances (Figure 2.13) and to dance faster (more 180 degree turns per minute) when they were foraging on higher-quality pollen. Once again the response showed a suprathreshold component and was demonstrated to be analog rather than binary.

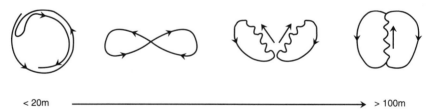

< 20m ————————————————————————→ > 100m

Figure 2.13. Honey bee foragers do recruitment dances on the vertical surface of a comb after they return from foraging trips. The shape of the dance changes with the distance to the source up to about 100 meters, depending on the geographic population of the bees. Typically they perform a round dance, left, to indicate a source close to the nest and a full-formed waggle dance to indicate a source about 100 meters or more distant. The arrows show the direction of body movement for the dances.

2.5.8 *Tasks Are Linked, Not Independent*

Worker honey bees change tasks as they age. These changes do not occur in fixed, temporally linear sequences but instead reflect changes in physiology of the bees and their location in the nest. There is a relationship between the performance of tasks and the location of bees in the nest. Bees emerge from their cells as adults near the middle of the nest, where the queen lays eggs in the individual cells of the combs, the eggs hatch, and younger nurse bees care for and feed the developing larvae. Tom Seeley grouped many of the tasks into four categories, or behavioral subcastes, that correlate with location in the nest. In his study, the youngest bees (0 to 2 days old) stayed near the center of the nest where they emerged and engaged primarily in cell-cleaning behavior. After two days, they moved to engage primarily in brood care, still in the central part of the nest; however, after about 11 days, they moved out of the brood nest and engaged in receiving, storing, and processing food. At the end of their third week of life they began foraging outside the nest. Second-order receiver bees (those that receive food from primary receiver bees) engaged in both food-storage and nurse bee activities.

2.5.9 *Foraging for Work, or Temporal Development?*

The centrifugal movement out of the center of the nest to the peripheral regions and the associated changes in stimuli encountered and tasks performed appear to be programmed developmental changes. They are not strictly age related because they can be retarded or accelerated with changes in the colony environment. An alternative hypothesis is that young bees get crowded out of the center of the nest by newly emerging workers and move to other areas of the nest and look for work. Foraging is a dangerous occupation and is performed by the oldest bees, so mortality is high among foragers and opens up vacancies in that task group. As a consequence, there is a constant flow of bees from the brood nest to the periphery and to the field.

The "foraging-for-work" hypothesis of Nigel Franks was tested by Nick Calderone. He placed brood combs in an incubator where the adults emerged over a 12-hour interval (Figure 2.14). Adult bees were marked with paint and put into two groups. One group went into a hive and experienced a normal environment as its members aged; members of the other group went into a cage with other newly emerged adult bees, were placed in an incubator, and were denied a normal social environment. Six days later another group of newly emerged bees was split, half going into the hive, the others into the incubator. Six days later (12 days after the first group), a final group emerged and was placed directly into the hive along with the 6- and 12-day incubator-treated bees. This resulted in three age groups of bees that experienced only natural hive conditions (0, 6, and 12 days) and two age groups that had been deprived of normal within-nest social experiences for 6 and 12 days, respectively. If the observed changes in task performance and nest location are consequences of centrifugal movement of bees, as proposed by the foraging-for-work hypothesis, then caged bees of different ages and the uncaged 0-day group should move through the nest and change tasks together. However, Calderone found that they differed in the tasks they performed, as one might expect for an internal, temporally correlated, physiologi-

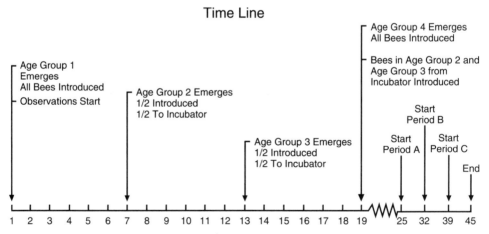

Figure 2.14. Time line for the temporal polyethism experiment. Reprinted from *Animal Behavior,* 51(3), Calderone NW, Page RE, "Temporal polyethism and behavioural canalization in the honey bee, *Apis mellifera,*" 631–643, Fig. 1 (1996), with permission from Elsevier.

cally based mechanism underlying changes in nest location and task performance.

Other studies likewise show that there are genetic and physiological factors that move worker honey bees through sets of tasks and hive location. Kim Fondrk and I selectively bred two distinct populations of bees that differ dramatically in the amount of pollen they store in the comb and in their foraging behavior (Chapter 5). We called them the high- and low-pollen-hoarding strains. When high- and low-strain workers are raised together in the same hive at the same time, the high-strain bees forage about 10 to 12 days earlier in life, pass through all the task transitions when they are younger, and spend less time engaged in each task (Adam Siegel and Robert Page, unpublished data). The transition from working in the nest to foraging is marked by a dramatic decrease in blood-circulating vitellogenin (a precursor of egg-yolk proteins in insects in general and a protein involved in the regulation of behavior in honey bees) and a concomitant increase in circulating juvenile hormone (Chapter 7). When bees are treated with an analog of juvenile hormone, they pass through the behavioral transitions faster and forage earlier in life. The same is true when the gene that makes vitellogenin is silenced. These results demonstrate that juvenile hormone and vitellogenin are physiological drivers of locational and behavioral changes.

We can conclude that temporally patterned physiological changes take place in workers as they age and move to different places in the nest, where they encounter and respond to different stimulus environments. In addition, their responsiveness to stimuli changes. We assume that they are "tuned" to the relevant stimuli in those locations and that the tuning subsequently affects division of labor. However, the rates at which they change location and tasks vary with the colony environment, and even the direction of change can be reversed.

2.5.10 Connectedness Is Not Constant, and K Does Not Equal N

Returning nectar foragers pass their loads to primary receivers, who in turn pass some of their load to secondary receivers (primary receivers may also place it in cells for processing and storage), who in turn pass the nectar to others in the nest. Therefore, each individual in the network of nectar transmission is a node for receiving and transmitting nectar and information. Dhurba Naug tracked the distribution of nectar in a hive and showed the asymmetries of distribution (Figure 2.15). Some individuals experience more receiving interactions than others, while others have more transmission interactions. He also showed that nectar flows through

Figure 2.15. Directed and weighted contact networks of interactions of worker honey bees in a nest. Each node (worker) is represented by a circle corresponding to a particular age class. Different shades of circles represent different age classes of bees. The connection between two nodes is represented by an arrow. The direction and thickness of the arrow correspond to the direction of flow and the total duration of contact, respectively. Note that the layout of the nodes does not represent their actual spatial locations in the colony. With kind permission from Springer Science+Business Media: "Structure of the social network and its influence on transmission dynamics in a honeybee colony," *Behavioral Ecology and Sociobiology* 62(11), 2008, 1719–1725, Naug D., Fig. 1.

the nest asymmetrically with respect to age and location of bees. Bees that are closer in age are more likely to engage in food exchange, and bees are more likely to exchange food with others nearby. The distribution of age in the nest, with the younger bees located centrally and older bees situated peripherally, sets up a centripetal flow of food from the older bees at the edges, including the foragers, to the younger bees in the center.

With the flow of nectar in the hive comes the flow of olfactory information and changes in response thresholds to sugar. Sucrose-response thresholds of hive bees are modulated by the concentration of sugar in solutions collected by foragers. Higher concentrations of solutions result in lower sucrose responsiveness of hive bees. Bees learn the odors of the foraging resources of the colony through associative processes during trophallaxis (the sharing of food by mouth-to-mouth exchange). Bees that have experienced interaction through food exchanges are more likely to engage in subsequent interactions with each other, creating a nonrandom network of interactions based on prior experience.

The example of nectar-flow dynamics shows direct effects on information flow and the decision functions of worker honey bees in the network. However, many, perhaps most, effects of the activities of bees are indirect, as in the pollen-foraging example in Figure 2.2. In that case, returning pollen foragers walk on the comb along the margins of the developing brood and the stored pollen. They take a statistical sample of the brood and stored pollen stimuli. They do not have complete information about the current need and supply, so they are not connected to all pollen foragers as indirect inhibitors of pollen foraging or the young larvae who are indirect releasers of pollen-foraging behavior—K does not equal N.

2.6 Summary Comments

The "spirit of the hive" referred to by Maurice Maeterlinck is the mechanisms through which coordinated activities can be performed by Charles Darwin's "crowd of bees working in a dark hive." The main theme of this chapter has been that the primary mechanism of social behavioral organization is the response of individual worker honey bees to their immediate environment and the consequential change in the stimulus. I presented a very simple "stone-soup" model showing that division of labor and what looks like coordinated behavior inescapably emerge from group living because individuals vary in their response sensitivities to stimuli or have different probabilities of encountering them. Then I presented an "avalanche of hard data," to borrow a phrase from the author of the foreward of this book, showing how the assumptions of the stone-soup model were wrong in details. But the model was useful for discussion of how we can get more complex behavior. Response thresholds can vary among individuals as a result of differences in individual experiences, differences in developmental states, or genetic variation. In Chapter 3, I will show how honey bee colonies contain a tremendous amount of genetic variation and how genotype affects the response thresholds and behavior of individual workers and colonies.

Suggested Reading

Calderone, N. W. 1995. Temporal division of labor in the honey bee, *Apis mellifera:* A developmental process or the result of environmental influences? *Canadian J. Zool.* 73:1410–1416.

Calderone, N. W., and Johnson, B. R. 2002. The within-nest behaviour of honeybee pollen foragers in colonies with a high or low need for pollen. *Anim. Behav.* 63:749–758.

Calderone, N. W., and Page, R. E. 1992. Effects of interactions among genotypically diverse nestmates on task specialization by foraging honeybees *(Apis mellifera). Behav. Ecol. Sociobiol.* 30:219–226.

Calderone, N. W., and Page, R. E. 1996. Temporal polyethism and behavioural canalization in the honey bee, *Apis mellifera. Anim. Behav.* 51:631–643.

Dreller, C., Page, R. E., and Fondrk, M. K. 1999. Regulation of pollen foraging in honeybee colonies: Effects of young brood, stored pollen, and empty space. *Behav. Ecol. Sociobiol.* 45:227–233.

Dreller, C., and Tarpy, D. R. 2000. Perception of the pollen need by foragers in a honeybee colony. *Anim. Behav.* 59:91–96.

Erber, J., Hoormann, J., and Scheiner, R. Phototactic behaviour correlates with gustatory responsiveness in honey bees *(Apis mellifera* L.). *Behav. Brain Res.* 174:174–180.

Farina, W. M., Grüter, C., Acosta, L., and McCabe, S. 2007. Honeybees learn floral odors while receiving nectar from foragers within the hive. *Naturwissenschaften* 94:55–60.

Fewell, J., and Page, R. E. 1999. The emergence of labor in forced associations of normally solitary ant queens. *Evol. Ecol. Res.* 1:537–548.

Frisch, K. von. 1967. *The Dance Language and Orientation of Bees.* Cambridge, MA: Belknap Press of Harvard University Press.

Gelfand, A. E., and Walker, C. C. 1984. *Ensemble Modeling.* New York: Marcel Dekker.

Goyret, J., and Farina, W. M. 2005. Non-random nectar unloading interactions between foragers and their receivers in the honeybee hive. *Naturwissenschaften* 92:440–443.

Grüter, C., and Farina, W. M. 2009. The honeybee waggle dance: Can we follow the steps? *Trends Ecol. Evol.* 24:242–247.

Harris, J. W., and Woodring, J. 1992. Effects of stress, age, season, and source colony on levels of octopamine, dopamine and serotonin in the honey bee *(Apis mellifera)* brain. *J. Insect. Physiol.* 38:29–35.

Haupt, S. S. 2004. Antennal sucrose perception in the honey bee (*Apis mellifera* L.): behavior and electrophysiology. *J. Comp. Physiol.* 190:735–745.

Humphries, M. A., Fondrk, M. K., and Page, R. E. 2005. Locomotion and the pollen-hoarding behavioral syndrome of the honey bee (*Apis mellifera* L.). *J. Comp. Physiol. A* 191:669–674.

Kaatz, H., Eichmüller, S., and Kreissl, S. 1994. Stimulatory effect of octopamine on juvenile hormone biosynthesis in honey bees *(Apis mellifera):* Physiological and immunocytochemical evidence. *J. Insect Physiol.* 40:865–872.

Kauffman, S. A. 1993. *The Origins of Order: Self-Organization and Selection in Evolution.* New York: Oxford University Press.

Martinez, A., and Farina, W. M. 2008. Honeybees modify gustatory responsiveness after receiving nectar from foragers within the hive. *Behav. Ecol. Sociobiol.* 62:529–535.

Naug, D. 2008. Structure of the social network and its influence on transmission dynamics in a honeybee colony. *Behav. Ecol. Sociobiol.* 62:1719–1725.

Pacheco, J., and Breed, M. D. 2008. Sucrose-response thresholds and the expression of behavioural tasks by middle-aged honeybee workers. *Anim. Behav.* 76:1641–1646.

Page, R. E. 2010. The "spirit of the hive" and how a superorganism evolves. In *Berlin-Brandenburgische Akademie der Wissenschaften Jahrbuch 2009.* Berlin: Akademie Verlag, pp. 513–532.

Page, R. E., and Erber, J. 2002. Levels of behavioral organization and the evolution of division of labor. *Naturwissenschaften* 89:91–106.

Page, R. E., and Mitchell, S. D. 1991. Self-organization and adaptation in insect societies. In *PSA 1990*, vol. 2, ed. A. Fine, M. Forbes, and L. Wessels. East Lansing, MI: Philosophy of Science Association, pp. 289–298.

Page, R. E., and Mitchell, S. D. 1998. Self-organization and the evolution of division of labor. *Apidologie* 29:171–190.

Page, R. E., Robinson, G. E., Britton, D. S., and Fondrk, M. K. 1992. Genotypic variability for rates of behavioral development in worker honeybees *(Apis mellifera). Behav. Ecol.* 4:173–180.

Page, R. E., Scheiner, R., Erber, J., and Amdam, G. V. 2006. The development and evolution of division of labor and foraging specialization in a social insect. *Curr. Top. Dev. Biol.* 74:253–286.

Pankiw, T., and Page, R. E. 2003. The effect of pheromones, hormones, and handling on sucrose response thresholds of honey bees (*Apis mellifera* L.). *J. Comp. Physiol. A* 189:675–684.

Pankiw, T., Page, R. E., and Fondrk, M. K. 1998. Brood pheromone stimulates pollen foraging in honey bees *(Apis mellifera). Behav. Ecol. Sociobiol.* 44:193–198.

Pankiw, T., Waddington, K. D., and Page, R. E. 2001. Modulation of sucrose response thresholds in honey bees (*Apis mellifera* L.): Influence of genotype, feeding, and foraging experience. *J. Comp. Physiol. A* 187:293–301.

Ramírez, G. P., Martínez, A. S., Fernández, V. M., Bielsa, G. C., et al. 2010. The influence of gustatory and olfactory experiences on responsiveness to reward in the honeybee. *PLoS ONE* 5:e13498. doi:10.1371/journal.pone .0013498.

Scheiner, R., Kuritz-Kaiser, A., Menzel, R., and Erber, J. 2005. Sensory responsiveness and the effects of equal subjective rewards on tactile learning and memory of honeybees. *Learn. Memory* 12:626–635.

Scheiner, R., Page, R. E., and Erber, J. 2004. Sucrose responsiveness and behavioral plasticity in honey bees. *Apidologie* 35:1–10.

Scheiner, R., Plückhahn, S., Öney, B., Blenau, W., et al. 2002. Behavioural pharmacology of octopamine, tyramine and dopamine in honey bees. *Behav. Brain Res.* 136:545–553.

Schulz, D. J., Sullivan, J. P., and Robinson, G. E. 2002. Juvenile hormone and octopamine in the regulation of division of labor in honey bee colonies. *Horm. Behav.* 42:222–231.

Seeley, T. D. 1982. Adaptive significance of the age polyethism schedule in honeybee colonies. *Behav. Ecol. Sociobiol.* 11:287–293.

Seeley, T. D. 1989. Social foraging in bees: How nectar foragers assess their colony's nutritional status. *Behav. Ecol. Sociobiol.* 24:181–199.

Seeley, T. D. 1995. *The Wisdom of the Hive: The Social Physiology of Honey Bee Colonies.* Cambridge, MA: Harvard University Press.

Siegel, A., Freedman, C., and Page, R. E. 2012. Ovarian control of nectar collection in the honey bee *(Apis mellifera). PLoS ONE* 7:e33465. doi:10.1371 /journal.pone.0033465.

Tofts, C., and Franks, N. R. 1992. Doing the right thing: Ants, honeybees and naked mole-rats. *Trends Ecol. Evol.* 7:346–349.

Tsuruda, J. M., and Page, R. E. 2009. The effects of foraging role and genotype on light and sucrose responsiveness in honey bees (*Apis mellifera* L.). *Behav. Brain Res.* 205:132–137.

Vaughan, D. M., and Calderone, N. W. 2002. Assessment of pollen stores by foragers in colonies of the honey bee, *Apis mellifera* L. *Insectes Soc.* 49:23–27.

Waddington, K. D. 1982. Honey bee foraging profitability and round dances. *J. Comp. Physiol.* 148:297–301.

Waddington, K. D., and Kirchner, W. H. 1992. Acoustical and behavioral correlates of profitability of food sources in honey bee round dances. *Ethology* 92:1–6.

Waddington, K. D., Nelson, C. M., and Page, R. E. 1998. Effects of pollen quality and genotype on the dance of foraging honey bees. *Anim. Behav.* 56:35–39.

— 3 —

Individual Variation in Behavior

A fundamental question in behavioral biology is "Why does animal A do X while animal B does Y?" The answer is usually found by looking for differences in the experiences of the individuals, the environments they occupy, or the genes and developmental processes that build their anatomical, physiological, and behavioral phenotypes. Before the mid-to late 1980s, observable variation in behavior among nestmate social insects was attributed almost exclusively to nongenetic factors. I remember that when I first encountered the book edited by Robert Jeanne in 1988 titled *Interindividual Behavioral Variability in Social Insects,* a collection of chapters written by key people in the field of social insect biology, I was struck by the lack of any consideration of the role of genetic variability within colonies. I decided that this needed to be investigated and set off on a quest that continues today.

In the sections that follow, I will present evidence demonstrating the effects of genetic variation on individual and colony behavior. First, I will show how the mating behavior of queens and the extremely high rates of genetic recombination during meiosis affect genetic variation within colonies. Then I will present evidence showing how nestmate workers with different fathers perform tasks with different probabilities and how behavioral interactions of individuals with different genotypes mutually affect their individual behavior, resulting in colony-level phenotypic effects. Finally, I will show the effects of an individual's genotype on behavioral plasticity at the individual and colony levels.

An explanation of terms is needed. A *gene* is a sequence of DNA along a chromosome that codes for a functional protein or peptide. A *genotype* is the combination of genes inherited from the mother and father in sexual diploid species. In practice, it can refer to either a genotype at a single gene locus or the entire set of genes of an individual. A *genome* a single set of chromosomes, as found in egg and sperm gametes, but the term has come to be used more broadly to refer to the entire DNA sequence of an individual or even a population, in the case of genome sequencing.

3.1 Genetic Variation and Behavior

Honey bee workers vary genetically for nearly every trait studied. Genetic variation affects the distribution of the stimulus-response relationships of the individuals in a nest, and social organization emerges through the fundamental self-organizing principles discussed in Chapter 2—the organizational and regulatory logic of the hive. The polyandrous mating behavior of the queen (queens mate with many males) and the high rate of genetic recombination affect the behavior of a colony because they affect the distribution of genetic variation underlying the response thresholds of workers. The genotype of an individual constrains its behavioral plasticity to environmental change and at the same time guards against the convergence of behavioral responses that could occur when individuals share a common environment and experiences. Individuals do still display plasticity in their behavior. Plasticity can arise through changes in the levels of behavior-releasing stimuli that are encountered or through the modification of the response thresholds of individuals as a consequence of experience or development. Some traits show considerably more plasticity at the individual level that in turn affects colony-level plasticity and resilience.

Plasticity and resilience need more explanation. Plasticity is the ability of an individual or a colony to change its behavior or social structure in response to an internal (physiological) or external (environmental) stimulus. For example, a colony increases its number of water foragers in response to an increase in hive temperature, while individuals respond by foraging for water and engaging in fanning behavior. Resilience is the ability to return to the previous colony or individual state when the stimulus returns to its previous level. For example, once the brood nest's temperature is reduced to about 35 degrees Celsius, water foraging ceases and individuals stop fanning.

3.2 Polyandry in the Honey Bee

Honey bee queens mate with a large number of males, and this polyandry produces a great amount of genetic variation among workers in a colony. This variation affects the distribution of responses to stimuli—the spirit of the hive—and colony behavior. Estimates of the number of matings based on genetic markers range from 7 to 18, depending on the population being measured. The average over all studies is about 12. The maximum number of confirmed matings, based on genetic markers of fathers contributing to the worker population of a colony, is 25. Queens make up to three mating flights within a few days after emerging as adults. On each flight a queen may not mate at all or may mate with up to 13 males. Mating flights usually take place between 1:00 p.m. and 5:00 p.m., a time of day when males are also flying. Adult males mature in the nest for about 5 days before they begin making daily flights, seeking out specific locations where drones from many colonies congregate and wait for queens. No one knows why certain locations are chosen year after year, or how the drones and queens find them. Norman Gary estimated that 25,000 drones from 200 colonies from within 5 kilometers attended one congregating area in Davis, California.

Before a queen enters a congregation site, the drones are flying around, apparently at random. However, when a queen enters, they quickly begin the chase and form comets consisting of hundreds of drones. The queen produces a chemical compound or pheromone, 9-oxo-2-decenoic acid, that is released from glands in her head and attracts the males. The males at the front of the comet scramble to reach the queen. The lucky ones mount her from above and behind and insert their aedeagus (the penis of a male insect) into the vaginal chamber of the queen. The aedeagus everts inside the sting chamber of the queen, and the male ejaculates his semen into the median oviduct (Figure 3.1). The semen is followed by a large mass of mucous that sits in the vaginal chamber and at the opening of the median oviduct. Then the male becomes paralyzed, falls back, the end of his aedeagus, the endophallus, breaks off, and he falls to the ground and dies. The next male mounts the female, displaces the previous male's mucous plug and endophallus, called the mating sign, and deposits his own semen. This takes place in rapid succession. The unlucky males fly home to their nests at the end of the afternoon and wait for the next day.

The queen returns to the hive with the mating sign of the last male still intact in her sting chamber, blocking the median oviduct and holding in the huge amount of semen deposited by the many male mates. The workers help the queen remove the mating sign, and over the next 24 hours the sperm migrate into the spermatheca, the sperm storage organ. The queen may make another flight the next day or several days later, but after this series of mating flights she will never again mate. She will store the sperm of her mates for her entire life, which can range from one to several years.

Until the early 1940s, the general belief was that queens mated with just one male. This was a logical assumption because of the presence of the mating sign, believed to be deposited by the male to block subsequent

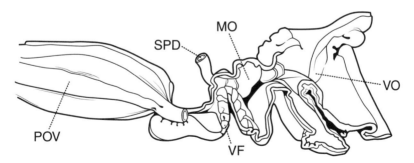

Figure 3.1. Reproductive tract of a honey bee queen. VO = vaginal orifice; MO = median oviduct; SPD = spermathecal duct; VF = valve fold; POV = paired oviducts. From Laidlaw, H. H. and R. E. Page, 1997, *Queen Rearing and Bee Breeding*, Fig. 77, Wicwas Press.

matings, and the fact that a queen stores roughly 5 to 6 million spermatozoa in her spermatheca, while individual males produce about 6 million. However, it was subsequently shown that queens mate with many
males. Therefore, a colony consists of a large number of subfamilies of
workers because potentially all the fathers contribute sperm to the eggs
of the queen (Figure 3.2). Worker honey bees with the same father share
75 percent of their genes in common (they are called super sisters because they inherit the entire genome of the father), while diploid species
share just 50 percent (full sisters). Half sisters of diploid species share
just 25 percent of their genes in common, the same as half sisters of
haplodiploid species.

Beginning in the 1950s, many believed that the sperm loads of the
different males clumped together in the spermatheca and were not uniformly distributed. Researchers thought that the sperm were used in
batches belonging to individual males. If this were true, it would reduce
the number of subfamilies in a colony and the amount of genetic variation at any given time. One argument for sperm clumping was based
on the observation that spermatozoa in a dissected, full spermatheca
appear to be arranged in "whorls" when they are viewed under a microscope, giving the appearance that they had entered in bundles, like
packages of spaghetti.

The other main piece of evidence used to support the sperm-clumping
idea was an article published in the mid-1950s by Stephen Taber. He
raised virgin queens that were homozygous for a recessive mutation,
called cordovan, that changes the black color of the bee to a light
brownish color. The head, antennae, thorax, and legs of honey bees are
normally black, but the abdomen may be striped with yellow and
black. Different strains of bees found in different parts of Europe, the
evolutionary home of the bees that were originally brought into North
America, vary in the amount of yellow found on the abdomen. He put
the queens in small hives, called mating nuclei, where they made mating flights and mated with the males from hives in the general vicinity.
Some of the males were sons of other queens that carried the mutation,
while others were wild type. The offspring of a cordovan male were easily distinguished by their color; those of wild-type males were black

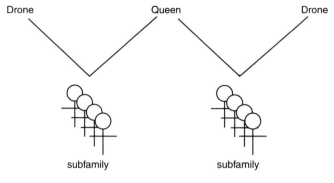

Figure 3.2. Subfamily diagram. The upper panel shows the subfamily composition of a colony derived from two drones mated to the same queen. The middle panel shows workers from a single colony where the queen was instrumentally inseminated with the sperm of six different males, each with different visible mutant markers. The workers of each subfamily are visibly distinguishable. The lower panel shows the workers that emerged as adults from a single comb from eggs laid by the queen. Workers from all six subfamilies are present in roughly equal proportions.

except for yellow stripes on the abdomen. He then sampled adult workers as they emerged from their cells on the comb over the course of the season. He found that the proportional representation (percentage) of the cordovan-type workers changed within individual colonies among samples. He concluded that the spermatozoa were not randomly or uniformly distributed in the spermatheca and that they must be clumping. However, both types were found in all colonies in every sample taken.

Several studies led to the decline of the sperm-clumping hypothesis. Ross Crozier and Dorothea Brückner published a note in the *American Naturalist* in 1981 pointing out that there was little evidence for sperm clumping in honey bees and that even Taber's data only suggested nonrandom fluctuations in sperm use over short time scales. Robert A. Metcalf and I had also come to that conclusion on the basis of Taber's results and in 1978 began to look at patterns of sperm use in colonies with naturally mated queens using allozyme markers. Allozymes are genetically different forms of enzymes that are easily detectable. The technique of isolating and visualizing allozymes was new and proving to be a great tool in studying and understanding genetic variation in populations of plants and animals. Metcalf was a pioneer in the use of these markers, publishing his first scientific paper in the major journal *Nature* while he was an undergraduate at the University of Illinois. He went on to do his doctorate with Edward O. Wilson at Harvard University, where he worked on a species of social wasps and was the first to directly measure genetic relatedness and test the inclusive-fitness theory of William D. Hamilton. We sampled emerging worker honey bees from two colonies over the course of 11 weeks and looked at the frequencies of three allozyme markers, the three forms of the enzyme malate dehydrogenase that we could unequivocally assign to fathers (Figure 3.3). However, we could distinguish only three groups of fathers that had common markers. Our results were similar to those of Taber: we found all three paternal allozyme markers in every sample. They fluctuated somewhat from sample to sample, but the sperm were not clumping together in a way that could significantly reduce the number

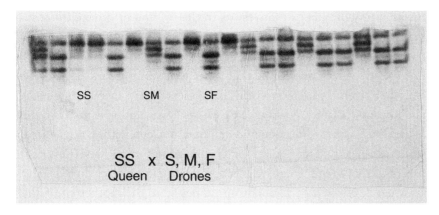

Figure 3.3. Acrylamide gel showing malate dehydrogenase allozyme phenotypes. The phenotype patterns correspond to three different genotypes of workers from a single hive. The queen was genetically homozygous for the slow (S) allozyme and was mated to one drone each having a slow, medium (M), or fast (F) allele. S, M, and F refer to how far the proteins travel on the gel. The proteins are seen stained. Each vertical lane contains the allozymes of a single individual. The lane marked SS contains the protein from an individual from the drone with the slow marker; SM had the medium father, and SF had a father with a fast allele.

of worker subfamilies. When two workers bump into each other in a honey bee colony, they are probably half sisters.

Robert Kimsey, Harry Laidlaw, and I looked at the filling of the spermatheca over time by instrumentally inseminating a series of queens and then dissecting the spermatheca from each queen at different times after insemination. We found that the spermatheca filled uniformly over time, becoming more densely packed as more spermatozoa entered it. They did not enter in packets. The whorls were formed by the spermatozoa after they had already entered the spermatheca.

3.3 Genetic Recombination in Honey Bees

Honey bees have adaptations that increase genetic variation within colonies: drone congregating areas where thousands of males compete for access to mating with queens, an extremely high number of mates, the anatomy of queens that results in the mixing of the sperm of her many mates, and an extremely high genetic recombination rate. Greg Hunt and I built the first genetic map for a social insect, one of the first for any insect, and found that the recombination rate of the honey bee was extremely high. It is still the highest of any animal mapped so far. Recombination occurs when gametes (eggs and sperm) are being made. Each gamete ultimately contains a nucleus with a single set of chromosomes that are parts of chromosomes inherited from both parents—they are recombined. Recombination takes place during meiosis when duplicated pairs of chromosomes from the mother and father come together, break in specific places, and reattach to form chromosomes that were derived from both parents (Figure 3.4). The frequency of breaking and recombining along a chromosome is called its recombination rate. The farther apart two parts of a chromosome are, the more likely it is that there will be recombination between them. Honey

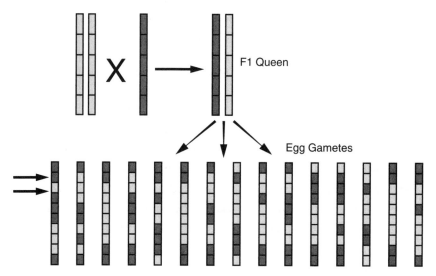

Figure 3.4. Recombination in a hybrid queen. A single chromosome is represented in this illustration. A queen from a "light gray" line was crossed to a haploid male from a "dark gray" line. They produced an F1 hybrid queen with one light gray and one dark gray chromosome. During oogenesis, the production of egg gametes, meiosis occurred where the chromosomes were broken and reassembled into a single chromosome that went into the egg.

bees have a very high rate of breaking and recombining along their chromosomes.

Recombination rates are measured in centimorgans (cM). One centimorgan distance between two places on a chromosome means that roughly 1 percent of the time there is recombination between those points. The distance between two points along a chromosome can be expressed as the physical distance measured by the number of nucleotide base pairs (bp), the building blocks of DNA, or the recombinational distance in cM. Another measure is the number of base pairs per cM of recombination. This is a way to compare genomes of different species. The honey bee has roughly 45,000 base pairs per cM. The mouse has about 2 million bp per cM, and humans have about 1 million bp per cM. The honey bee has about 20 times more recombination than humans.

Recombination breaks up gene combinations inherited from individual parents and builds new combinations in the gametes, thereby increasing the genotypic variation of the offspring produced. Therefore, it appears that genotypic variation is very important for honey bees. Why? There are two ways to approach this question. One is to look at the consequences of genotypic variation on individual and colony behavior. The other is to ask whether higher genotypic diversity results in higher fitness for colonies, higher survival and reproductive success (successful queens and drones), and the evolution of higher rates of recombination and polyandry. I have tried to keep these questions separate. In the rest of this chapter, I will discuss genetic variation for individual behavior and its consequences. In particular, I will discuss how the genotype of a worker places constraints on her behavioral flexibility (plasticity), and the effects of genetic variation and polyandry on colony plasticity and resilience, the ability to respond to environmental changes and return to "normal." In Chapter 4, I will discuss the evolution of polyandry.

3.4 Genetic Variation Is Necessary for Evolution

Genetic variation is the grist of the mill of natural selection, grinding out evolutionary change. Evolution is a process. Within a population there is variation for phenotypic traits. A phenotype is what you see. It can be simple, such as eye color, or complex, like an entire eye. It is a result of the genes and regulatory networks that affect the trait, the developmental processes that build it, and the environment, a partner with the genome in the determination of the phenotype. Natural selection is simply the consequence of differential reproductive success. Individuals with some traits survive and reproduce more than those with alternative traits. If there is a correlation in a trait between parents and offspring (the trait is heritable), then the favored trait will increase in representation in the next generation. It evolves.

Without genetic variation for traits affecting social organization, there can be no evolution of colony traits such as nest architecture, food storage, reproduction, and defense. Variation can affect social signaling networks, such as pheromones produced by workers and queens or the stimulus-response systems of individuals.

3.5 Genetic Variation for Worker Behavior

In 1986, Nick Calderone, Kim Fondrk, Gene Robinson, and I began looking at the effects of genetic variation on individual worker behavior within colonies. Calderone was my first doctoral student, Robinson was the first postdoctoral fellow in my lab, and Fondrk was my first technician. When we began our experiments, the general belief in the social insect community was that genetic variation among workers within a colony had no effect on individual behavior and that observed variation was a consequence of experience and colony need.

Genetic variation in a population is distributed within colonies and among colonies. If individuals within colonies are highly related, as when a queen mates with one male, then the genetic variation within colonies is relatively low, and variation among colonies is relatively high. If, however, relatedness within colonies is low, as when queens mate with many males, genetic variation within colonies increases, but it decreases among colonies. This has profound effects on how we study within-colony variation in behavior and the response of populations to selection on colonies. We need genetic variation within colonies to study the effects of genotypes on behavior, but we need to limit within-colony genetic variation to breed successfully for traits. Populations evolve faster with more variation among colonies.

Rick Hellmich and Walter Rothenbuhler controlled mating and selected bees for the amount of surplus pollen they store (pollen hoarding). I succeeded Rothenbuhler at The Ohio State University (OSU) when he retired, and had access to the strains. The high-pollen-hoarding strain stored large quantities of pollen, while the low-pollen-hoarding strain stored less. Nick Calderone reared bees from the OSU high and low strains together in the same colonies and found that the high-strain bees foraged earlier in life and were more likely to collect pollen than were the bees from the low strain. This was a clear demonstration of a genetic component of interindividual variation in behavior among workers within a colony. In this case, however, the bees were not true sisters but were derived from different artificially selected populations and were adopted into colonies.

Gene Robinson and I looked at variation among sisters within colonies. We took advantage of the subfamily composition of honey bee colonies (Figure 3.3) to study the effects of genotype on individual behavior. Workers belonging to the same subfamily are more similar in their genotypes and behavior than are members of different subfamilies. We instrumentally inseminated queens with known allozyme markers for the metabolic enzyme malate dehydrogenase with three males, each with a distinct marker (Figure 3.3). We could distinguish the father of all the workers on the basis of the allozyme markers they inherited. All workers were raised in the same colonies at the same time and shared the queen as their mother.

We collected bees engaged in collecting nectar, collecting pollen, guarding the entrance of the nest, removing dead bodies (undertaking), and performing recruitment dances on the surface of a swarm (scouting). Nectar and pollen foragers were collected at the entrance as they returned. They were classified as having a pollen load, an easily observable character, or having no pollen. This was a crude distinction because pollen foragers may also have nectar, and those with no pollen could have contained water or been empty. Guards stand at the entrance of the nest with an easily identifiable posture, wings held out, mandibles open, and forelegs lifted, and check the credentials (nest membership) of individuals entering the nest by touching them with their antennae. Undertakers remove dead bodies from the nest. Some bees die inside the nest, but most die in the field. We placed dead workers in the hive, waited for undertaker bees to pull them out of the hive entrance, and collected the undertakers. Scouts on swarms were collected by turning the test colonies into artificial swarms using a technique I learned from Norman Gary. All the bees were shaken from the hive into a box, and the queen was caged. The bees and the queen were placed in a cool, dark place overnight and fed sucrose solution. The next day, the queen was suspended in her cage approximately 1.25 meters off the ground, and bees were shaken from the box. The workers took flight, detected the queen's pheromones, and settled on her in a swarm. Within a short time, some bees initiated searches for new homes and began performing recruitment dances providing distance and directional information to the nest sites they had discovered. We collected

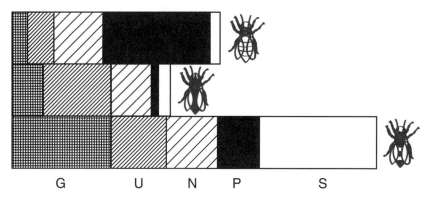

G   U   N   P   S

Figure 3.5. Relative proportion of bees of different genotypes (different patterns of workers) that were collected engaged in guarding (G), undertaking (U), collecting nectar (N), collecting pollen (P), or scouting on swarms (S).

the dancing bees. All bees were processed, analyzed for allozyme markers, and assigned to colony subfamilies. We found striking differences in the behavioral profiles for workers from different paternal subfamilies (Figure 3.5).

In another series of experiments, Claudia Dreller, a postdoctoral researcher in my lab, sampled returning foragers from a natural nest of feral honey bees. Using a set of genetic markers, she found that as in the previous study, pollen and nectar foragers differed with respect to the representation of the different subfamilies of the colony, thus demonstrating the same phenomenon in a naturally occurring nest with a naturally mated queen.

3.6 Behavioral Plasticity and Constraints

Nick Calderone also performed a cross-fostering experiment with the high- and low-pollen-hoarding strains produced by Hellmich and Rothenbuhler. He placed newly emerged high- and low-strain workers in high- and low-strain colonies and waited for them to forage. Foraging behavior differed for high- and low-strain bees in both colony environments. High-strain bees were always more likely to collect pollen. High- and low-strain bees also differed in their behavior depending on whether they were in high- or low-strain colonies. This demonstrated that they were responding to stimuli in the nest. High-strain bees increased their pollen foraging in low-strain colonies, while low-strain bees decreased their pollen foraging in high-strain colonies, showing behavioral plasticity. This is also a demonstration of the amplification of differences in behavior of bees of different genetic makeup due to differences in the stimulus environment. Amplification increases specialization among nestmates. In this case, differences in the nest environments were consequences of the genotypic composition of the colonies—the "nature of nurture." Differences between high- and low-strain foragers were much greater in the high-strain colonies, where high-strain bees were 21 times more likely to return with a load of pollen, than when they were raised in low-strain colonies, where they were only 2 times more likely to return with a pollen load.

Although high- and low-strain bees demonstrated behavioral plasticity, they were also constrained by their genotypes. Extreme changes in social environment, such as those that occurred between the high- and low-strain colonies, did not make the behavior of the high- and low-strain workers the same. This can also be demonstrated by testing bees derived from the high- and low-pollen-hoarding strains, or European (EHB) and Africanized (AHB) honey bees, for their responses to stimulation with sugar using the proboscis extension response assay (Figure 2.7). (Honey bees are not native to the New World. They were initially imported from Europe into North and South America, and today commercial beekeepers in the United States still use European honey bees. However, honey bees were imported from Africa into Brazil in the mid-1950s and hybridized with the European bees that were there. They subsequently spread across South America, through Central America and Mexico, and into the southwestern United States. They are known as Africanized honey bees.) If bees are of approximately the same age and have been maintained under approximately the same conditions, then high-strain bees are more responsive than low-strain bees, and AHB are more responsive to sucrose solutions than EHB. Responses modulate with experience, environment, and age, but the range over which they change is constrained to a measurable extent by the genotype of the bee.

3.7 Genetic and Behavioral Dominance

Genetic and behavioral dominance can also occur. Genetic dominance occurs when one form (allele) of a given gene has more of an effect on a phenotype than another allele when they both occur in an individual— the individual is heterozygous for the gene. This is the case for eye color. If you inherit a blue allele from one parent and a brown allele from the other, you most likely have brown eyes because the brown allele demonstrates genetic dominance over the blue. Your eye color more often is not an average of the blue and brown phenotypes. In bees, wild-type black cuticle color (+) is dominant over the cordovan (cd). A +/cd heterozygote is black.

Behavioral dominance occurs when a colony phenotype does not reflect the average phenotypic value of the genotypes of individual worker residents but is overly influenced by one genotypic group. Nest defense demonstrates both genetic and behavioral dominance. Africanized honey bees have been shown to respond much faster to stimuli that release stinging behavior, and many more members of the colony respond. My former doctoral student Ernesto Guzmán-Novoa set up colonies in Mexico that consisted of mixtures of Africanized and European honey bees. He instrumentally inseminated European queens with sperm from European and Africanized drones to produce colonies that were composed of European and hybrid workers. By varying the proportions of the AHB and EHB semen used for the inseminations, he produced colonies that contained 0 percent, 25 percent, and 50 percent hybrid workers (Figure 3.6). He also had the colony that contributed the AHB drones for insemination (the drone father colony) and other

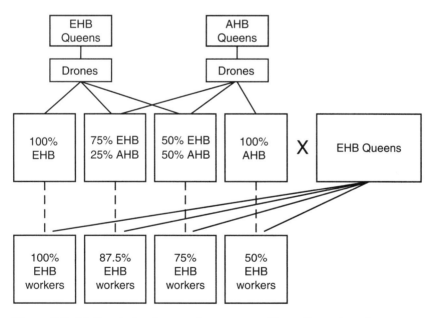

Figure 3.6. Mating design for experiment to test effects of genetic and behavioral dominance on defensive behavior.

captured, feral AHB colonies for comparisons. Defensive behavior was measured by waving a 5-by-6-centimeter black leather patch at the end of a meter stick in front of the entrance. The time it took for the first bee to sting the patch and the total number of stingers in the patch after one minute were recorded (Figure 3.7). Worker honey bees have barbed stingers and leave the sting apparatus and their intestines behind when they sting, the ultimate self-sacrifice. Dark colors and movement are both stimuli that release stinging behavior in bees.

Africanized colonies (100 percent AHB) responded eight times faster and deposited more than five times as many stings as European colonies. This was consistent with the findings of many other studies. We would expect the hybrid colonies where workers had EHB mothers and AHB fathers (50 percent Africanized) to be intermediate in response, but the hybrid colonies gave nearly the same defensive behavior response as AHB colonies, demonstrating genetic dominance for the AHB high defensive behavior trait. Times to sting the patch for the colonies with 25 percent hybrid bees were significantly faster than expected on the basis of genetic dominance and the genotypic mixture of bees, demonstrating behavioral dominance for that trait. Stingers were removed from the patches and analyzed for enzyme markers. The EHB and AHB fathers used to instrumentally inseminate the EHB queens had distinct genetic (allozyme) markers. We could distinguish which stingers came from EHB and which came from EHB X AHB hybrid crosses. Africanized hybrid stingers were always overrepresented relative to the proportional representation of workers in the hive, which suggests that the original owners of the stingers had lower response thresholds to the dark color and movement. As the proportion of AHB increased in hives, the number of stingers increased, and more EHB were stimulated to sting. When a worker stings, she releases an alarm pheromone composed primarily of isopentyl acetate. This compound smells like banana oil and is a strong releaser of stinging behavior. By stinging the patch, the bees altered the stimulus environment and the behavior of their nestmates, making them more likely to respond and sting—a correlation between the response and the stimulus environment.

Figure 3.7. Upper panel: Testing an Africanized honey bee colony for defensive behavior by waiving a black leather patch in front of the entrance. Middle panel: A leather patch full of stingers following a defensive behavior test. Lower panel: The hand of one of the operators who chose not to wear protective gloves. Photo by Ernesto Guzmán-Novoa.

The experiment discussed in the preceding paragraph only suggested that the hybrid bees were more responsive to stimuli that release stinging behavior. An alternative explanation is that they were more likely to be guard bees at the entrance, and, therefore, more of them encountered sting-inducing stimuli. Guard bees sit at the entrance and respond to stimuli, releasing an alarm pheromone that recruits bees engaged in other activities into the task of colony defense. We collected guards at the entrance to see whether AHB X EHB hybrids were more likely to guard. We found no difference in the likelihood that an EHB or a hybrid performed guard duty. Hence, they differed in their responses to the sting-releasing stimuli.

3.8 Behavioral Plasticity and Colony Resilience

Colonies need to be able to respond to changes in the environment. Behavioral flexibility of individual bees leads to the resilience of colonies, the ability to respond and return to a "normal" state. So how plastic is behavior, and how resilient are colonies to perturbation?

3.8.1 Thermal Regulation

Gene Robinson and I proposed that colony plasticity and resilience could occur as a consequence of polyandry and genetic variation of response thresholds of individuals. A broad distribution of response thresholds might result in more appropriate, graded responses of colonies to perturbations of the stimulus environment (Figures 2.4C and 2.4D). Bees regulate the temperature of the brood nest where the eggs, larvae, and pupae reside. When the temperature exceeds about 35 degrees Celsius, the bees collect water and initiate fanning, setting up an evaporative cooler. When the temperature falls below 35 degrees Celsius, they cluster around the brood, providing insulation, and vibrate their wing muscles and generate heat. If a colony is subjected to an increasing ambient temperature, rather than all individuals immediately initiating foraging for water and ventilating, it might be best if those with the lowest response thresholds to elevated temperatures turn on first, followed by those with progressively higher thresholds.

Theoretical models and studies of thermoregulation in colonies with more or less genetic variation due to different numbers of subfamilies have validated the graded-response hypothesis. Colonies with multiple subfamilies of workers (more fathers) are better able to maintain a constant temperature and respond better to increasing temperatures (are more resilient) than are colonies with a single father because individuals with increasing response thresholds to temperature respond and fan as the temperature increases, yielding a gradual response of the colony as a whole.

3.8.2 Stored Pollen

Colonies show great resilience to disturbance of stored pollen quantities. When stored pollen is added to a colony, foragers collect less pollen until the added surplus is consumed. If stored pollen is removed, foragers collect more pollen until its level is restored. This occurs through individuals increasing their foraging effort for pollen by collecting larger loads and making more frequent trips (individual plasticity) and through the recruitment of new individuals into pollen foraging (colony plasticity). A low pollen-foraging stimulus results in a higher degree of differentiation among subfamilies, where fewer subfamilies collect larger shares of the incoming pollen. A high stimulus results in the recruitment of new pollen foragers from across a broader representation of subfamilies (like fanning behavior with high temperatures discussed in Section 3.8.1); it exceeds the response thresholds of a broader group of individuals. These results are consistent with the response-threshold model for colony plasticity and resilience.

3.8.3 Undertaking Behavior

Undertaking behavior does not demonstrate the same level of individual and colony plasticity or resilience. Gene Robinson and I tested the behavioral plasticity of workers and colony resilience in response to increased bee mortality. Most bees die while foraging away from the nest. However, Norman Gary estimated that about 10 to 15 percent of bees die in the nest. A colony during the peak period of summer will produce about 1,500 new bees each day, with an equal number dying from old age, which in a bee is usually associated with wearing out as a forager. Therefore, roughly 150 to 200 bees die each day inside the nest. Occasionally colonies encounter pathogens or natural or human-made toxic substances collected in nectar or pollen that are returned to the hive and create higher mortality. These bodies must be removed, or the nest space will fill up with corpses, and the bodies will breed diseases. Within a colony there are bees called undertakers that specialize in removing the dead bodies. They drag the dead bees out the entrance and then take off and carry them through the air, dropping them up to a few hundred meters away (Figure 3.8).

We increased the number of dead bees in colonies by putting 15 freeze-killed bees in the hive every 15 minutes. We collected the first 50 undertakers, our low-stimulus sample. We then dumped 1,000 dead bees into

Figure 3.8. Undertaker bees removing a dead body from the hive. Photo by Jacob Sahertian.

hives and collected all the undertakers. The colonies consisted of three subfamilies of bees (derived from three fathers used to instrumentally inseminate the queens) that were identifiable with allozyme markers. All undertakers were captured as they dragged bodies out of the hive, and their subfamily membership was determined by allozyme analysis. Our prediction was that some subfamilies would be more likely than others to remove the dead bees; that is, they would have genotypes that made them more responsive to the dead-bee stimulus. This prediction was confirmed. However, we predicted that with 1,000 dead bees, the response thresholds of bees across all subfamilies would be exceeded, and the biases observed with the low stimulus would be eliminated. That was not the case. The same biases were observed with the high stimulus. The workers did not show plasticity for undertaking; instead, they were "constrained" by their genes. We also failed to see plasticity at the colony level. Repeated sampling depleted the undertakers in the hive such that the time to remove the dead bees increased significantly, and it took several days to return to the initial rates of removal. This suggested that there was a finite number of undertakers composed of a nonrandom genetic sample of bees.

3.8.4 Canalization of Behavior by the Environment

Bees that live together in a nest share common colony environments and experiences. If behavioral responses are "tuned" by the environment or the "needs of the colony," then the outcome will be a decrease in variation among workers for their response thresholds, measured by their response probabilities. The common environment will make them more similar; it will canalize behavior. The genotype of an individual places constraints on her plasticity and maintains differences among workers in genetically variable colonies. Nick Calderone denied bees early social experiences by putting them in cages in an incubator for 6 or 12 days (Figure 2.14). He recombined them with bees from the same age cohorts that had resided in the hive and compared their behavior. The bees were from the high- and low-pollen-hoarding strains of Hellmich and Rothenbuhler and were known to differ in their foraging behavior. High-strain bees that had been deprived of a social environment were still more likely to collect pollen than were low-strain bees that had been deprived or not deprived. This showed that the genotypes of the bees constrained them from environmental canalization of their foraging behavior.

Stored pollen acts as an inhibitor of pollen foraging; young larvae stimulate pollen foraging. We can manipulate the stimulus environment of hives by varying these two hive components. A colony with more stored pollen and less brood collects less pollen than a comparable colony that has been manipulated to have less stored pollen and more young larvae. Tanya Pankiw set up colonies that were manipulated to have a high-pollen-foraging-stimulus environment and paired them with colonies that had a low-pollen-foraging-stimulus environment. She placed marked, same-aged groups of bees derived from the high- and low-pollen-hoarding strains of Page and Fondrk (Chapter 5) and waited for them to forage. When they foraged, she collected them and weighed their nectar and pollen loads. She found that high-strain bees were more likely to collect larger loads of pollen and smaller loads of nectar than were low-strain bees, regardless of the hive's stimulus environment, and both strains collected more pollen under the high stimulus. High- and low-strain genotypes responded equally to changes in the pollen-foraging stimuli, but behavioral differences were maintained across all tested environments because of the constraints placed on behavior by the genotypes of the bees.

3.9 Laying-Worker Behavior

The ovaries of a worker honey bee will develop in the absence of the queen and larvae. Both produce pheromones that suppress ovary development and egg-laying behavior. After the queen dies, bees attempt to raise a new queen from larvae. They get just one opportunity over a very few days to select larvae, begin feeding them in excess to trigger the alternative developmental programs that produce queens rather than workers, and raise one successfully to maturity. If they fail, they become hopelessly queenless, and the colony begins its development into a terminal state of being a laying-worker colony. Laying-worker colonies of temperate populations of bees begin with a few bees engaging in egg laying, usually beginning within 12 to 28 days after losing the queen. The eggs are very quickly removed and consumed by bees that inspect cells. Eggs laid by workers are in some way distinguishable from eggs laid by queens and are eaten by nurse bees. Initially, no eggs survive to hatch. Then over time, more bees begin laying eggs and fewer bees engage in egg eating, the egg eaters are probably becoming egg layers, and some larvae are reared into mature drones. As time passes, more bees lay eggs and fewer forage, and eventually chaos reigns where the larvae die, the adults age and die, and the colony dies. However, a colony may produce up to 6,000 drones before it succumbs, providing opportunities for workers to have some reproductive success if any of the males get lucky.

There is evidence in the form of caste-specific reproductive behavior that workers do produce successful sons. Honey bees construct three different types of cells for brood rearing: small worker-sized cells, somewhat larger drone-sized cells, and large queen cells. Queens control the fertilization of eggs when they lay in worker- and drone-sized cells to produce either females or males. Unmated virgin queens show a bias for laying eggs in worker-sized cells; they do not "know" that they are not inseminated. Unfertilized eggs laid in worker-sized cells produce smaller drones, about half the normal size, with reduced numbers of spermatozoa, and probably are not competitive for mating with queens. Workers are not mated, have no sperm to fertilize eggs, and show a significant bias for laying eggs in drone-sized cells, which result in full-sized, fully functional males. Therefore, the egg-laying behavior of unfertilized workers, producing competitive sons, is probably a caste-specific adaptation that could not have evolved without the reproductive success of the drones produced by workers.

Gene Robinson and I showed that workers with different fathers within the same queenless colony differ in their likelihoods of laying eggs and engaging in oophagy (the consumption of eggs laid by workers). There is genetic variation. Workers show different physiological and behavioral responses to the absence of the pheromonal inhibitors of producing and laying eggs. This is probably a consequence of having different rates of developmental change associated with ovary development and the associated changes in stimulus-response relationships. These differences at the individual level affect the overall production of males by laying workers, a colony-level reproductive trait of queenless colonies that is an aggregate of the individual success of laying workers and conflict among them. Colonies with some combinations of genotypes very quickly begin producing males and produce many. Others maintain too few egg layers and too many egg eaters for too long and fail to produce any males.

3.10 Summary Comments

In this chapter I have presented evidence for effects of genetic variation on within-nest variation in behavior. I have also presented colony-level consequences of genetic variation for worker behavior in the form of effects on plasticity, resilience, and behavioral dominance. In the next chapter, I will address the evolution of polyandry and the maintenance of behavioral genetic variation.

Suggested Reading

Beye, M., Gattermeier, I., Hasselmann, J. M., Gempe, T., et al. 2006. Exceptionally high levels of recombination across the honey bee genome. *Genome Res.* 16:1339–1344.

Calderone, N. W., and Page, R. E. 1988. Genotypic variability in age polyethism and task specialization in the honey bee, *Apis mellifera* (Hymenoptera: Apidae). *Behav. Ecol. Sociobiol.* 22:17–25.

Calderone, N. W., and Page, R. E. 1992. Effects of interactions among genotypically diverse nestmates on task specialization by foraging honeybees *(Apis mellifera). Behav. Ecol. Sociobiol.* 30:219–226.

Calderone, N. W., Robinson, G. E., and Page, R. E. 1989. Genetic structure and division of labor in honeybee societies. *Experientia* 45:765–767.

Collins, A. M., Rinderer, T. E., Harbo, J. R., and Bolton, A. B. 1982. Colony defense by Africanized and European honey bees. *Science* 218:72–74.

Crozier, R. H., and Brückner, D. 1981. Sperm clumping and the population genetics of Hymenoptera. *Am. Nat.* 117:561–563.

Crozier, R. H., and Page, R. E. 1985. On being the right size: Male contributions and multiple mating in social Hymenoptera. *Behav. Ecol. Sociobiol.* 18:105–115.

Dreller, C., Fondrk, M. K., and Page, R. E. 1995. Genetic variability affects the behavior of foragers in a feral honeybee colony. *Naturwissenschaften* 82:243–245.

Fewell, J. H., and Page, R. E. 1993. Genotypic variation in foraging responses to environmental stimuli by honey bees, *Apis mellifera. Experientia* 49:1106–1112.

Fewell, J. H., and Page, R. E. 2000. Colony-level selection effects on individual and colony foraging task performance in honey bees, *Apis mellifera* L. *Behav. Ecol. Sociobiol.* 48:173–181.

Fewell, J. H., and Winston, M. L. 1992. Colony state and regulation of pollen foraging in the honey bee, *Apis mellifera* L. *Behav. Ecol. Sociobiol.* 30:387–393.

Guzmán-Novoa, E., Hunt, G. J., Page, R. E., and Fondrk, M. K. 2002. Genetic correlations among honey bee (*Apis mellifera* L.) behavioral characteristics and body size. *Ann. Entomol. Soc. Am.* 95:402–406.

Guzmán-Novoa, E., and Page, R. E. 1994. Genetic dominance and worker interactions affect honey bee colony defense. *Behav. Ecol.* 5:91–97.

Guzmán-Novoa, E., Page, R. E., and Correa-Benitez, A. 1997. Introduction and acceptance of European queens in Africanized and European honey bee (*Apis mellifera* L.) colonies. *Am. Bee J.* 137:667–668.

Hamilton, W. D. 1972. Altruism and related phenomena, mainly in social insects. *Ann. Rev. Ecol. Syst.* 3:193–232.

Hellmich, R. L., Kulincevic, J. M., and Rothenbuhler, W. C. 1985. Selection for high and low pollen-hoarding honey bees. *J. Hered.* 76:155–158.

Hunt, G. J., and Page, R. E. 1995. A linkage map of the honey bee, *Apis mellifera*, based on RAPD markers. *Genetics* 139:1371–1382.

Jeanne, R. L., ed. 1988. *Interindividual Behavioral Variability in Social Insects.* Boulder, CO: Westview Press.

Jones, J. C., and Oldroyd, B. P. 2007. Nest thermoregulation in social insects. *Adv. Insect Physiol.* 33:153–191.

Laidlaw, H. H., and Page, R. E. 1984. Polyandry in honey bees: Sperm utilization and intra-colony genetic relationships. *Genetics* 108:985–997.

Laidlaw, H. H., and Page, R. E. 1997. *Queen Rearing and Bee Breeding.* Cheshire, CT: Wicwas Press.

Myerscough, M. R., and Oldroyd, B. P. 2004. Simulation models of the role of genetic variability in social insect task allocation. *Insect Soc.* 51:146–152.

Page, R. E. 1986. Sperm utilization by social insects. *Ann. Rev. Entomol.* 31:297–320.

Page, R. E., and Erickson, E. H. 1988. Reproduction by worker honey bees. *Behav. Ecol. Sociobiol.* 23:117–126.

Page, R. E., Kimsey, R. B., and Laidlaw, H. H. 1984. Migration and dispersal of spermatozoa in spermathecae of queen honey bees (*Apis mellifera* L.). *Experientia* 40:182–184.

Page, R. E., and Laidlaw, H. H. 1988. Full sisters and super sisters: A terminological paradigm. *Anim. Behav.* 36:944–945.

Page, R. E., and Metcalf, R. A. 1982. Multiple mating, sperm utilization, and social evolution. *Am. Nat.* 119:263–281.

Page, R. E., and Mitchell, S. D. 1998. Self-organization and the evolution of division of labor. *Apidologie* 29:171–190.

Page, R. E., and Peng, C. Y.-S. 2001. Aging and development in social insects with emphasis on the honey bee, *Apis mellifera* L. *Exper. Gerontol.* 36:695–711.

Page, R. E., and Robinson, G. E. 1994. Reproductive competition in queenless honey bee colonies (*Apis mellifera* L.). *Behav. Ecol. Sociobiol.* 35:99–107.

Pankiw, T., and Page, R. E. 1999. The effect of genotype, age, sex, and caste on response thresholds to sucrose and foraging behavior of honey bees (*Apis mellifera* L.). *J. Comp. Physiol. A* 185:207–213.

Pankiw, T., and Page, R. E. 2001. Genotype and colony environment affect honey bee (*Apis mellifera* L.) development and foraging behavior. *Behav. Ecol. Sociobiol.* 51:87–94.

Robinson, G. E., and Page, R. E. 1989. Genetic basis for division of labor in insect societies. In *The Genetics of Social Evolution*, ed. M. D. Breed and R. E. Page. Boulder, CO: Westview Press, pp. 61–80.

Robinson, G. E., and Page, R. E. 1995. Genotypic constraints for corpse removal in honey bee colonies. *Anim. Behav.* 49:867–876.

Scheiner, R., Page, R. E., and Erber, J. 2004. Sucrose responsiveness and behavioral plasticity in honey bees. *Apidologie* 35:1–10.

Schlüns, H., Schlüns, E. A., van Praagh, J., and Moritz, R. F. A. 2003. Sperm numbers in drone honeybees *(Apis mellifera)* depend on body size. *Apidologie* 34:577–584.

Taber, S. 1955. Sperm distribution in the spermathecae of multiple-mated queen honey bees. *J. Econ. Entomol.* 48:522–525.

Tarpy, D. R., and Nielsen, D. I. 2002. Sampling error, effective paternity, and estimating the genetic structure of honey bee colonies (Hymenoptera: Apidae). *Ann. Entomol. Soc. Am.* 95:513–528.

Visscher, P. K. 1983. The honey bee way of death: Necrophoric behavior in *Apis mellifera*. *Anim. Behav.* 31:1070–1076.

Winston, M. L. 1987. *The Biology of the Honey Bee*. Cambridge, MA: Harvard University Press.

— 4 —

The Evolution of Polyandry

Honey bee queens mate with a large number of males—they are polyandrous. Estimates of the number of mates range up to more than 20. Understanding polyandry is important for understanding the mechanisms of social behavior (the spirit of the hive) because there are phenotypic consequences of genetic variation at the individual and colony levels. But we also would like to understand its evolution, and it is important to keep the questions of consequences and evolution separate. Chapter 3 dealt with the behavioral consequences of within-colony genetic variation for behavior, some of which is a consequence of polyandry, some a consequence of genetic recombination. In this chapter, I discuss its evolution. I begin by introducing hypotheses that have been proposed and what I think are the necessary components and requirements of a sufficient evolutionary model. I then present in detail one model that I think fulfills the requirements, the sex-determination hypothesis, followed by a discussion and evaluation of other hypotheses that provide plausible explanations.

4.1 Why Do Queens Mate with So Many Males?

The question of the evolution of polyandry was coincidental with the realization that queen honey bees mate with many males and mix their sperm. Until then it had been generally accepted that the queen mated so many times because she needed all those sperm, up to many times the amount one male produces, to fill her spermatheca. However, she stores only roughly the amount produced by one male, so why is she so inefficient in acquiring sperm? She has what appear to be clear anatomical adaptations to be inefficient. (As a queen flies through the air, each of her mates deposits his semen into the median oviduct [see Figure 3.1]. Each load of semen pushes the previous deposits back into the lateral oviducts, which are fluted and capable of expanding like large balloons. After returning to the nest, the queen sits around for up to 24 hours pulsating her abdomen, squeezing the oviducts like toothpaste tubes, and sending the sperm back past the spermathecal duct and into the vaginal orifice, where it is then expelled in thin threads that the workers remove. The spermatozoa of each male have access to the spermatheca only while they sit at the opening of the spermathecal duct, where they can then actively swim up the duct and into the spermatheca. By regulating the process of passing sperm back, the queen regulates the mixing of the sperm of her many mates.) And why don't males make more sperm, enough to fill the spermatheca? Another explanation was that queens simply cannot resist the mating attempts of males as they fly through the drone congregating areas. Multiple mating is just a matter of convenience; the effort for queens to resist the males is too costly. Then why do they make additional mating flights after they have mated with enough males to fill their spermatheca?

There have been many hypotheses about the evolution of polyandry in honey bees and other social insects since the publication of the first detailed model of polyandry in 1980. Currently, there are at least 14 recognized hypotheses, which fall into two categories: those that are about genetic heterogeneity and those that are not. The sperm-need and mating-convenience hypotheses discussed in the preceding paragraph are not about genetic heterogeneity and are not supported by critical analyses. Most reviews of polyandry models consider that only three of the remaining genetic-heterogeneity hypotheses warrant further study: sex determination, parasites and pathogens, and division of labor. I will discuss them in Sections 4.2, 4.3, and 4.4. In my view, the evolution of polyandry is really a two-part problem, and a sufficient model must address both. First is the question of what maintains genetic variation in the population. This is not trivial because natural selection acts to reduce genetic variation. If a trait is very important for an organism, and one heritable phenotypic variant is better with respect to survival and reproduction, the less favorable variant will eventually be eliminated or driven to very low frequencies in the population, taking with it the genetic variation for the trait. Any genetic-variation-based model of the evolution of polyandry must have a mechanism to maintain genetic variation against directional selection. Second, it must show how increasing numbers of mates affects the genotypic composition of colonies and colony fitness (survival and reproduction).

4.2 Sex Determination and Polyandry

The sex-determination hypothesis for the evolution of polyandry is, in my opinion, a front-runner for honey bees and also for some other social Hymenoptera. It proposes that polyandry in honey bees may be explained by the effects of the sex-determination system on colony survival and reproductive success. Honey bees are haplodiploid. Males have just one set of chromosomes, while females have two. Because males and females have different numbers of chromosomes, there must be a genetic mechanism whereby female development is initiated in diploids and male development in haploids. The mechanism involves a single gene called the complementary sex-determining *(csd)* gene. It is also called the X locus and the sex locus. Different forms of this gene are called sex alleles.

4.2.1 Complementary Sex Determination

In 1845, Johannes Dzierson, a parish priest in Silesia, now a province of Poland, first proposed that drone honey bees have mothers but no fathers. This was the first proposal for a sex-determination mechanism for any animal. However, the complementary sex-determining genic mechanism was not proposed for honey bees until 100 years later by Otto Mackensen following its discovery by Phineas Westcott (P. W.) Whiting in *Bracon hebetor,* a small parasitic wasp. Whiting showed that to be a female, one must inherit two different forms (alleles) of the *csd* gene; one must be heterozygous. Individuals that have two copies of the same allele are homozygous; haploid individuals have just one copy, are hemizygous, and are effectively homozygous. Homozygous and hemizygous individuals develop into males (Figure 4.1). Diploid *B. hebetor* males are sterile.

With the development of instrumental insemination technology in the late 1940s and early 1950s came the ability to control the mating of honey bee queens. Early honey bee geneticists wanted to use the inbred-hybrid method of breeding that was so successful in the production of hybrid corn. Maternal lines of honey bees were inbred for a few genera-

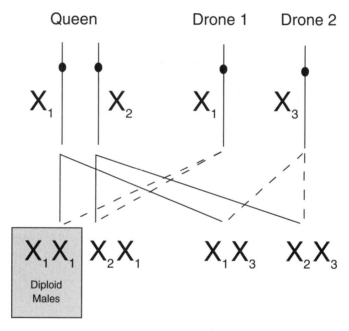

Figure 4.1. Diagram of sex determination in the honey bee. Solid lines represent egg gametes; dashed lines represent sperm. The queen mated with two males, one having a sex allele identical to one of hers, the other different. Only the chromosome containing the *csd* locus is represented. Half the offspring derived from drone 1 are homozygous and develop into diploid males. Normal females have two different alleles, and normal males have just one set of chromosomes. From Laidlaw, H. H. and R. E. Page. 1997. *Queen Rearing and Bee Breeding*, Fig. 93. Wicwas Press.

tions, and then inbred lines were crossed to look for line combinations that combined well and produced superior stock. Immediately it became obvious that honey bees suffered severely from inbreeding; the viability of the larvae decreased to where half the larvae died. Otto Mackensen, a bee researcher for the U.S. Department of Agriculture, immediately suspected that this must be related to the complementary sex determination found by Whiting in *B. hebetor*. He proposed that homozygosity in bees was lethal rather than producing sterile diploid males, as occurred in *Bracon*. Jerzy Woyke, a Polish bee researcher and geneticist, performed some elegant experiments and showed that in fact the homozygous diploid male larvae do not die from the effects of lethal genes. Instead, these larvae are eaten by nurse bees during their first few hours of life. They produce some kind of "eat me" cue that results in their removal from the cell in which they are developing (Figure 4.2).

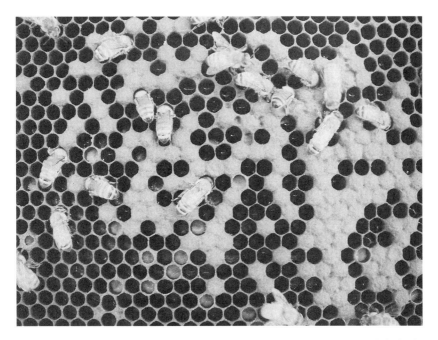

Figure 4.2. Frame containing capped brood. The empty cells contained diploid male larvae that were removed by workers. This pattern is called *shot brood*. Photo by Kim Fondrk.

He was able to rear the newly hatched larvae in the laboratory for their first 72 hours of life and return them to their cells, where the nurse bees cared for them as if they were normal. He produced adult diploid males and showed that they were effectively sterile—evolutionary dead ends. Then over a period of 13 years (from 1990 to 2003) the *csd* gene was genetically mapped, isolated, and sequenced, and its complementary mode of action was confirmed by students and postdocs in my laboratory and the laboratory of my former postdoc, Martin Beye.

4.2.2 *Population Genetics of Sex Determination*

Homozygosity at the sex locus results in the death of larvae due to sibicide, sisters killing sisters in this case. So how common is homozygosity? The population genetics of *csd* have been carefully explored. Honey bee populations have between 6 and 18 sex alleles that are functionally complementary, depending on the population measured. The number of sex alleles is a consequence of selection-mutation-drift equilibrium. New alleles arise through mutation, a random change in the DNA sequence associated with the region of the gene responsible for complementarity. Beye and members of his lab have been characterizing the nature of the alleles and the kinds of mutations that can occur and be compatible with other alleles. The fate of most new mutational variants of most genes is to be eliminated from the population soon after they occur because they reduce the survival and reproductive success of their owners or are lost to genetic drift.

Genetic drift is simply sampling error. Imagine that you have a bowl with 100 randomly mixed marbles. Half the marbles are blue; the others are red. You draw exactly 4 marbles from the bowl. The chances are that frequently you will not get exactly 2 of each color. Let us say that you get 3 of one color and 1 of the other, an occurrence with relatively high expectation (50 percent chance). In this case, let us make it 3 reds and 1 blue. You empty the bowl and now fill it with the proportional colors that you drew, the same as if each marble made 25 new copies of itself to go into the bowl. Now the bowl has 75 reds and 25 blues. The proportion of reds and blues has drifted. The next time you draw 4, the chances are quite good that you will get 4 reds (32 percent chance). When you reconstitute your population of marbles for the next bowl generation, you will no longer have blue marbles. They will have been lost because of sampling error, that is, drift.

This same process goes on in real honey bee populations from generation to generation. The proportional representation of the sex alleles drifts because there is not an exactly equal representation of all of them from one generation to the next. Some colonies contribute more gametes to the mating pool by making more reproductive offspring (queens and drones); others fail to survive because of chance occurrences, like disease or predation—events that may have nothing to do with which sex alleles they have. The more marbles you sample from the bowl each generation to constitute the next, the longer it will take to lose one of the colors, and the same holds for losing sex alleles in populations. The larger the breeding population (queens and successful drones), the longer sex alleles stay in the population, and, therefore, more sex alleles persist. With some assumptions about the size of the breeding population and mutation rates (how often a mutation occurs at the *csd* locus), population geneticists can predict how many alleles are expected in different populations.

Another prediction from population genetics is that the sex alleles will be nearly equal in frequency in the population. Natural selection maintains alleles in populations. This is a consequence of their complementarity. Only diploid heterozygotes can survive. Because eggs and sperm gametes fuse at fertilization to make a zygote, gametes with rare sex alleles are less likely to pair with alleles that are like them than are gametes with common alleles. Therefore, rare alleles will be favored (fewer homozygous combinations), and common alleles will be selected against (more homozygous combinations). At equilibrium, if a very large population is assumed, all alleles will be equally frequent. If there are 10 alleles in the population, each will represent 10 percent of the genes in diploid females as a consequence of pairing of sperm and egg gametes, and also in the males as a consequence of being derived directly from the gametes of the heterozygous queens.

Imagine that there are p sex alleles in a population, each at equal frequency. Then the frequency of each will be $1/p$ at equilibrium. What is the probability that a queen will mate with a male having a sex allele identical to one of hers? She has two different alleles, so for each male with which she mates, the chance that he will have one like hers is $2/p$. For example, if there are 10 sex alleles in the population, each at a frequency of 0.10 (10 percent), then she has a 20 percent likelihood $(2 \times 0.1 = 0.20)$ of mating with a male that has an allele like one of hers every time she mates. So, on average 20 percent of the males she mates with will have a matching allele. If she mates with 10 males, on average two will have an allele that matches hers. Half of the offspring derived from those two males will be diploid males and will not survive. The average proportion of diploid larvae produced by that queen will be 0.20 (the proportion of males with matching alleles) $\times$ 0.50 (the proportion of the offspring of that male that will be homozygous; remember that the queen has two different alleles) $= 0.10$. Her expected brood viability then is 0.90 or 90 percent.

If queens in a population mate just one time, then there will be only two classes of queens, those that produce 50 percent diploid males (50 percent viable brood) from their fertilized eggs and those that produce none (100 percent viable brood). If there are 10 sex alleles at equal frequency, then 80 percent of the queens will have 100 percent brood viability, and 20 percent will have 50 percent. If a queen mates a very large number of times, then the representation of sex alleles in her spermatheca, which are derived from her mates, will get very close to that of the whole population. So, a queen that mates with a very large number of males in our population with 10 sex alleles will have 20 percent of the sperm in her spermatheca with sex alleles that match an allele in the eggs she produces, resulting in diploid males. In this case all queens that mate this very large number of times will have the same brood viability, 90 percent. You can see that the number of times a queen mates affects the distribution of brood viability among queens in a population (Figure 4.3). But you can also see that the average brood viability for all queens in a population remains the same regardless of the number of times queens mate. In our example population with single mating, the

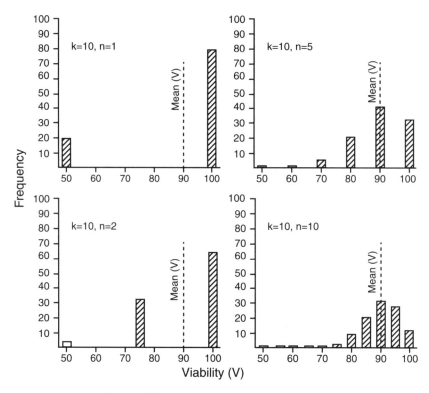

Figure 4.3. Distribution of brood viability among queens of a population based on the number of times they mate (*n*). It is assumed that there are 10 sex alleles at equal frequencies in the population. Reprinted from Page and Metcalf 1982. Multiple mating, sperm utilization, and social evolution. *Am. Nat.* 119:263–281, Fig. 2, with permission from University of Chicago Press.

average brood viability is 90 percent because 20 percent of the queens have half their diploid larvae homozygous, resulting in 10 percent diploid males, and 80 percent of the queens have 100 percent female brood, so $(0.2 \times 0.5) + (0.80 \times 1.00) = 0.90$ (90 percent).

4.2.3 Mating and Fitness

So far, I have shown a mechanism for maintaining genetic variation in the population against directional selection, the requirement that females have two different sex alleles. I have also shown how the number of mates affects the distribution of alleles and brood viability in the population. But to have a complete model, I need to show how brood viability affects the fitness of queens that mate with different numbers of males. Imagine that colonies survive and reproduce in proportion to the viability of their brood. In that case, queens that mate one time will have on average a fitness of 0.80 relative to the maximum of 1.00. Queens that mate a very large number of times will also have a relative fitness of 0.80, so there is no advantage for any genotypes that affect number of mates. However, what if only colonies with more than 75 percent viable brood are able to produce worker populations that are sufficient to maintain the colony, survive, and produce successful new queens and drones? In that case queens that mate just one time will have a 20 percent chance of failing, while those that mate a very large number of times will be successful, a clear advantage for genotypes that result in polyandrous mating. This model can be developed more completely to show different kinds of relationships between the relative fitness of mating number genotypes and brood viability, but the example just shown is sufficient (Figure 4.4). In general, as long as the relationship between fitness (including survival and reproduction) and brood viability is concave, multiple mating will be selected.

Dave Tarpy, a former graduate student of mine, tested the sex-determination model for the evolution of polyandry. He instrumentally inseminated 31 queens, each with the semen of three different drones that were brothers of the queens. The expected result of the inbreeding was that the matings would produce an average of 25 percent diploid progeny, or a brood viability of 75 percent. The actual median for the 31 queens was 72 percent. Each male produces a different amount of sperm, and because brother drones were selected as fathers at random, there was a continuous distribution of brood viability between about 50 percent and 100 percent (Figure 4.4). Colonies were allowed to develop through-out the season and to acquire resources necessary for surviving the winter. All colonies with at least the median of 72 percent viable brood survived the winter and built back up in the spring, while 63 percent of the colonies with less than the median viable brood perished. He also estimated worker populations in the colonies, the most important de-

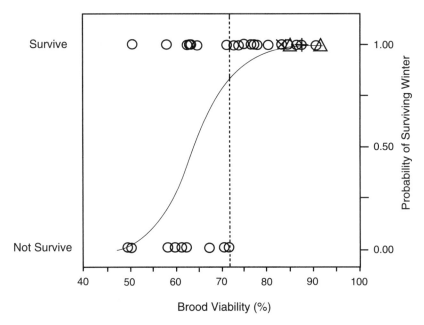

Figure 4.4. Survival of colonies over the winter was affected by brood viability resulting from homozygosity at the sex locus. The sigmoid curve is the relationship between brood viability and the probability of surviving the winter. Note that the curve is sigmoid, not linear, and that the average brood viability is in the concave part of the curve. This is important for polyandry to evolve under the sex-determination model for the evolution of polyandry. Reprinted with kind permission from Springer Science+Business Media: *Behavioral Ecology and Sociobiology,* "Sex determination and the evolution of polyandry in honey bees *(Apis mellifera),*" 52(2), 2002, 143–150, Tarpy DR, Page RE, Fig. 1.

terminant of survival and reproductive output of colonies, and found a concave relationship between brood viability and number of workers. This test validated the sex-determination hypothesis as one plausible explanation for the evolution of polyandry in honey bees.

4.2.4 Difficulties with the Hypothesis

However, there are difficulties with the hypothesis. First, it would be nice to have a single hypothesis that can also explain the extraordinary recombination rate in the honey bee (Section 3.3)—a single hypothesis that explains all the mechanisms for increasing genotypic variation. But the high recombination rate cannot be explained on the basis of the *csd* alone because *csd* is a single gene and recombination has no effect on it. There must be something else going on as well.

Another problem is that an increase in the number of mates has a big effect on the variance in brood viability in populations up to about 6 to 8 mates but is negligible above 10. So, if we assume that mating is costly to queens in terms of their potential survival and reproductive success, how can we explain mating numbers that average more than 10 in some populations? There are at least two possible explanations: (1) the fitness costs of mating are low; and (2) queens do not count the number of matings and, therefore, do not have much control.

4.2.5 Is There a Fitness Cost to Mating?

The model presented in Section 4.2.3 asks only whether more mates result in higher fitness than fewer mates. However, there must be costs of mating that will ultimately limit the number. At a minimum, queens that mate with too many males may be injured by the overexpansion of their reproductive tract from the semen they receive. If there is an advantage to polyandry, then the number of mates should reach a level where the benefits of mating for the survival and reproductive success of the queen equal the costs of mating. If the costs are high, then queens should mate fewer times. Mating costs of queen honey bees have been assumed to be high. Queens can be eaten by predators on flights, fail to find their way back to their nest, or contract sexually transmitted diseases from their mates. Dave Tarpy estimated the costs of mating from a study where 32 queens made a total of 60 flights, including orientation flights. He timed the flights to obtain the total number of minutes of flight and, using genetic markers, determined the number of mates of each queen. He also recorded the number of queens that failed to return. He argued that the costs of mating flights should be put in terms of the number of mating flights or minutes of flight rather than number of mates because failure to find the nest and exposure to predators are more significant components of mating risk. Two of 32 queens

(6.25 percent) failed to return, 0.0333 per flight, 0.0026 per minute of flight, or 0.0125 per mating. He concluded that this is a relatively low risk.

4.2.6 Can Queens Count?

Can queens actually count the number of mates on a given mating flight? This is an important question if queens are in some way optimizing the number of times they mate by balancing costs and benefits. Estimates of the number of mates of individual queens show tremendous range and variation. If queens are regulating the number of mates, why is there so much variance? Dave Tarpy let 30 queens make one mating flight and then restricted them from making a second flight. Eight queens, however, attempted a second flight and were caught in a trap as they tried to leave the hive. Using molecular markers, he compared the number of times queens that attempted a second flight mated with the number of matings of those that attempted just one mating flight. He found no difference in the average number of mates; they were both about 5. One queen had 0 successful matings on her first flight and did not attempt a second. The number of times queens mated ranged from 0 to 13 for a single flight. He also compared the amount of time they spent on their mating flights and found no correlation between numbers of mates and how long they flew. The conclusion was that queens just make flights. Some queens fly for longer times, perhaps because they fly to drone congregations farther away or have difficulty finding them. When they fly through the congregation areas, they mate with males without assessing number and fly home. Some make a second flight, but very few make a third. The net result is a great range in the number of mates and a huge amount of variation, not expectations for a trait that is being regulated. A more likely explanation is that queens make a sufficient number of flights and fly for long-enough times to get enough mates to make it very likely that their brood viability is above the survival threshold.

One caveat is needed. Jerzy Woyke did show a correlation between the amount of sperm in the spermathecae of queens that made one mating flight and those that made two. He split the group that made two flights into those that were successful on their second flight and those that were not. Interestingly, there was an inexplicable difference between those that were successful on their second flight and those that failed to mate on the second flight. The ones that failed had more spermatozoa from the first flight than those that succeeded. Overall, the number of spermatozoa stored by queens from their first flight, and made a second flight, successful or not, ranged from 0.287 million to 6.96 million sperm. Queens that made just one mating flight and did not attempt another had 1.7 to 7.2 million. These are tremendous ranges that suggest only weak regulation of sperm intake of queens, at best. He did not determine the number of times the queens mated, which is the most important measure for all genetic-diversity hypotheses.

4.3 Pathogens and Parasites

Another plausible hypothesis for the evolution of polyandry in honey bees proposes that polyandry evolved as a mechanism to resist pathogens and parasites. There are two variations on this hypothesis: the Red Queen and the general-resistance hypotheses.

4.3.1 *The Red Queen*

The Red Queen in Lewis Carroll's *Through the Looking Glass* said to Alice as she was running and going nowhere, "Now here, you see, it takes all the running you can do, to keep in the same place." The Red Queen hypothesis for evolutionary biology was originally coined by Leigh Van Valen as a metaphor for evolutionary processes in constantly changing environments. A population may constantly be evolving useful traits to deal with some particular environment, including predators, pathogens, and parasites, but the environment itself is changing, so the population is never able to evolve the optimum phenotype. It keeps evolving toward something, but it never gets there. This metaphor has been used to explain the evolution of polyandry in honey bees. The argument is similar to one used to explain the evolution of sex as a mechanism to deal with parasites and pathogens.

Pathogens and parasites have very short generation times (hours to days or weeks) compared with their hosts (months or years for honey bees, depending on whether they are tropical or temperate). As a consequence, pathogens and parasites evolve faster than their hosts and put the hosts in the position of always evolving resistance (running), which the pathogens and parasites are always evolving mechanisms to overcome. The model assumes that resistance in the host is based on a single gene that is matched by a gene in the pathogen that lets it invade and spread. Gene-for-gene resistance-interaction complexes have been shown for plant-insect-pathogen systems. Polyandry and genetic recombination generate variation among the larvae in a nest. A pathogen may get into a colony and into a larval host but not have the appropriate genotype to develop well in that specific larva. The gene in the pathogen does not match the gene in the larva. But because the pathogen has a very short generation time, it will not take long before a mutation or a recombinant type occurs with a genotype that is very well suited to that particular larval genotype. The pathogen then increases in the larva, and its offspring spread to the neighboring larvae. If the next larva it spreads to has the same genotype as the original host larva, it will reproduce rapidly in that larva and spread to the next, and so on, eventually weakening or killing all the larvae and the colony, as frequently happens with American foulbrood disease (Section 4.3.2). However, with genetic recombination and polyandry, the probability increases that the genotypes of the larvae nearby are different from that of the current larval hosts. When the pathogen spreads to a neighbor, it will not be well adapted to it, the larva initially will have some degree of resistance, and the spread of the disease will be slower. The pathogen will need to adapt to every new host it invades.

The result is that genetic variation will be maintained in the population for the genes engaged in the gene-for-gene arms race with the pathogen because there is no optimal or best genotype. The same is true for the pathogen: there is no optimal genotype. So this model satisfies the criterion of a mechanism to maintain variation (presented in Section 4.2). Its action is the same as that presented for *csd* and the sex-determination model. Selection is on the variance, not the mean effects.

There will always be susceptible and resistant genotypes in colonies. Absolute resistance is not achieved. An increase in the numbers of mates decreases the variance in disease load of colonies just as it decreases the variance in diploid males and brood inviability. If there is a concave relationship between disease load and colony reproductive output (successful queens and drones), then polyandry will be favored. Or if there is a threshold effect where colonies below a certain disease load die without reproducing, polyandry will be selected.

How plausible is this model for explaining polyandry in honey bees? Genetic variation for disease resistance has been demonstrated repeatedly. However, there is little evidence for the gene-for-gene mechanisms. Gene-for-gene resistance could be demonstrated if different subfamilies within a colony showed differential infection from the same strain of pathogen. Harry Laidlaw instrumentally inseminated a queen with the semen of six different males. The queens were homozygous for two visible recessive mutations, one that affects integument color (cordovan, *cd*), the other eye color (snow, *s*). As discussed in Section 3.2, individuals homozygous for the recessive cordovan allele have a brown cuticle where bees are normally black. There are three alleles for the snow eye-color gene: wild type (+), snow *(s)*, and tan *(s^t)*. The wild type is dominant, so any individual with a wild-type allele will have black eyes. Individuals homozygous (or hemizygous in the case of males) for snow have white eyes, while individuals that get a white allele paired with a tan allele have red eyes. Cordovan workers with red eyes are striking in appearance. Homozygous workers or hemizygous males with tan alleles have tan eyes. We used males that were either wild type or cordovan for integument color and were wild-type, snow, or tan eyed. Workers derived from the six combinations had either cordovan or wild-type integument and wild-type, snow, or red eyes.

Queens were placed in colonies and began to lay eggs. The experiment was originally designed to study patterns of sperm use, another look at sperm clumping and sperm competition, so emerging worker samples were taken periodically over about three months. During the sampling, one colony developed American foulbrood disease, a serious disease of larvae caused by the bacterium *Paenibacillus larvae*. Associated with this was the complete loss of one of the six subfamilies from samples (a presumably susceptible subfamily) and a huge increase in the proportion of another subfamily relative to the rest (resistant). The colony was treated with antibiotics to kill the *P. larvae,* and subfamily proportions returned to where they had been before infection. Although these results are anecdotal, they provide support for genetic variation in larval susceptibility to disease, an expectation of a gene-for-gene mechanism of resistance, but they do not rule out a polygenic (multiple-gene) mechanism. Kellie Palmer and Ben Oldroyd did a similar experiment where they inoculated colonies with multiply and singly inseminated queens with the same strain of *P. larvae* and had similar, but again not conclusive, results.

4.3.2 General Resistance

The most complete and best-known studies of honey bee disease resistance are those of Walter Rothenbuhler. Rothenbuhler worked with an American-foulbrood-resistant strain of bees derived from the apiary of a beekeeper, Harold Brown. American foulbrood is one of the most lethal honey bee diseases and continuously plagues commercial beekeeping. Effective antibiotic chemical treatments exist, and today most beekeepers depend on them. However, Brown did not treat his colonies but let them die of the disease, and eventually the surviving generations were resistant. Rothenbuhler studied the mechanisms of resistance and found that they were behavioral, anatomical, and physiological.

Brown's bees displayed the hygienic behavior of uncapping cells with dead pupae and removing them from the nest. This is a trait that is found at low frequencies in most populations but had been selected to a high frequency in his bees. Removing the dead pupae and scales, the spore-laden residue of the dead bees, reduced the amount of inoculum in the hive and greatly reduced levels of infection. Hygienic behavior has also been shown to be effective against chalkbrood, a fungal disease caused by *Ascosphaera apis.* Brown's bees also were better able to filter out bacterial spores from food, presumably because they had a more efficient filtering system on the proventricular valve, a constriction with filtering hairs that lies between the nectar stomach (crop) and the midgut (Figure 2.8).

Two physiological mechanisms were found. Only small larvae are susceptible to invasion by the bacterium. Larvae heavier than about 0.7 milligrams are resistant. Brown's larvae at first developed faster, which allowed them to exceed this minimum quickly, and then slowed development so that they were no larger at the end of larval development. The brood food produced by Brown's nurse bees contained an antibacterial substance that made bacterial spores less likely to germinate and also had a bactericidal effect on vegetative cells.

Tom Seeley and Dave Tarpy tested colonies with multiply and singly inseminated queens after they had been inoculated with *P. larvae* spores. Colonies with multiply inseminated queens had more brood and larger adult populations and were heavier with stored food than those with queens inseminated by one male. They also had reduced disease intensity, the number of diseased larvae per comb of brood. Unfortunately, this measure confounds the effects of polyandry on brood area with the disease susceptibility of the larvae. The spread of the disease could have been equal in the two types of colonies, but the colonies with more brood distributed over more combs, perhaps a consequence of polyandry, would have had lower disease intensities even though they had the same number of diseased larvae. The multiply inseminated group had significantly less variance in disease, which supports the hypothesis that the trait variation is a consequence of genetic variation, but not the parasite-pathogen hypothesis for the evolution of polyandry.

Genetic variation was also shown for chalkbrood disease. Inoculated colonies with multiply inseminated queens had reduced variance in the prevalence of the disease but did not differ in their averages. In addition, colonies were tested for hygienic behavior by using standard methods. Colonies with multiply inseminated queens had less variance in hygienic behavior, although the means did not differ between groups. An additional study using the same methods, but without inoculating with pathogens, showed no effect of number of queen mates on the prevalence of sacbrood (a viral disease), American foulbrood, or Euro-

pean foulbrood (a bacterial disease), but did show a significance difference in disease intensity for chalkbrood (a fungal disease).

From the combined studies just described we can conclude that there
is selectable genetic variation for many different mechanisms of resistance to disease. Selection in the case of American foulbrood resulted
in changes in a suite of traits involving behavior, anatomy, and physiology. Genetic-diversity studies have shown that multiple mating decreases disease prevalence for American foulbrood, but the mechanism
is unclear. The prevalance of chalkbrood disease is reduced at some
times but not at others, and the variance is lower when colonies have
more genotypic diversity (more drone fathers). Multiple mating reduces the variance among colonies in hygienic behavior, a behavioral
mechanism with broad effects on disease resistance. Hygienic behavior
reduces chalkbrood prevalence.

Although there is little support for the Red Queen hypothesis, and
equivocal support for the general-resistance hypothesis, there is clear
support for hygienic behavior as a mechanism to reduce disease by
eliminating the sources of inoculum, dead larvae and pupae. So why is
hygienic behavior rare? I assume that there must be a cost to the behavior that outweighs its benefit for disease resistance, but I can only spec-
ulate what that cost might be (Section 4.4.2). The effect of the cost is to
make alleles for the behavior relatively rare and effective in low doses in
colonies. In this case, polyandry could increase the chance that a queen
mates with a few males, not too many, that carry the alleles for hygienic
behavior. With behavioral dominance, the colony could be hygienic
with a small percentage of workers engaged at a reduced cost—a
genotypic-diversity hypothesis.

4.4 Genotypic Diversity and Division of Labor

Division-of-labor hypotheses for the evolution of polyandry come in
two forms: *genetic specialists* (genotypic variation results in more
specialists and higher-performing colonies) and *colony stability* (genotypic variation buffers colonies from the effects of changes in the
environment).

4.4.1 Genetic Specialists

The genetic-specialist hypothesis proposes that polyandry evolved as a mechanism to get a broader distribution of worker genotypes in colonies and thus broader distributions of response thresholds. As a consequence, individuals are self-sorted into performing those tasks for which they have the lowest response thresholds. Individuals with lower thresholds will be more likely to perform the tasks repeatedly as specialists and will become more efficient. Increased efficiency is believed to be the reason division of labor and task specialization evolved in social insects.

The genetic-specialist hypothesis assumes that there is genetic variation for performing tasks, that there is a mechanism (deterministic or stochastic) for maintaining genetic variation for task performance, that genetic specialists are more efficient, and that colonies with more genetic specialists survive and reproduce more than those with fewer specialists. It is clear from many studies that the genotype of an individual biases her behavior toward differentially performing some tasks. Within-colony genetic variation has been demonstrated for foraging for pollen, nectar, and water, dancing on a swarm, guarding the entrance of the nest, stinging behavior, undertaking, provisioning queen versus worker cells, the age of onset of foraging, precocious foraging (initiating foraging behavior very early in life in the absence of older bees), fanning, worker egg-laying behavior, eating worker-laid eggs, grooming, and engaging in food exchange. There is also some experimental evidence that polyandry can lead to faster rates of colony growth, better survival, and production of more males (a component of reproduction) than colonies with single-mated queens. However, we still lack a sufficient model to explain the maintenance of genetic variation for task performance and a clear demonstration of increased efficiency in genetic specialists.

4.4.1.1 A Model for the Maintenance of Variation and the Evolution of Polyandry   Clearly, genotypic variability and polyandry are not necessary for division of labor within colonies because single individuals change tasks in response to changes in age and colony need, for example, changing the loading biases for pollen and nectar in response to changes in colony stimuli. Colonies with once-mated queens construct comb, raise brood, and forage for pollen and nectar; in short, they appear to perform all the normal functions of colonies, even after strong selection for single behavioral traits. It is unlikely that a specific genotype is needed in any colony to ensure that any normal task is performed. However, a primary condition for the maintenance of genetic variability and for the evolution of polyandry as a consequence of genotypic effects on task specialization is that individuals vary in their value as specialists as a consequence of their genotypes. For example, individuals with genotype A are more likely to perform task T_A and do it better than individuals of genotype B, while individuals with genotype B are more likely to perform T_B and do it better. There also cannot be a single genotype that does both T_A and T_B as well as the individual specialist genotypes.

Assume that the decision to forage for pollen and nectar is controlled by two independent genes (loci), say, *P1* and *P2*. Each locus has two alleles, *h* (high) and *l* (low), where the *h* allele results in a bias toward collecting pollen and the *l* allele results in a bias toward nectar. The loci and alleles are additive and equal in their effects. Also assume that the action of the two loci is based on the ratio of high, *h*, and low, *l*, "doses" at both loci. For example, individuals with the genotype $P1_h P1_h P2_h P2_h$ have four doses of high alleles and will show a stronger tendency to collect pollen than those with a $P1_l P1_l P2_l P2_l$ genotype. Individuals with $P1_h P1_h P2_l P2_l$ or $P1_h P1_l P2_h P2_l$ genotypes will be equivalent and intermediate. In order for this model to work, genotypes are not additive across individuals. For example, a colony composed of equal numbers of foragers with a $P1_l P1_l P2_l P2_l$ genotype and foragers with a $P1_h P1_h P2_h P2_h$ genotype would be better (closer to the optimum uptake of pollen and nectar) than a colony that is composed entirely of $P1_h P1_h P2_l P2_l$, $P1_h P1_l P2_h P2_l$, or $P1_l P1_l P2_h P2_h$ genotypes, even though there is the same dose of high alleles in each of the colonies. The reason they are different is the increase in efficiency due to specialization, another assumption of the model. If these assumptions are met, there is no single best

Table 4.1 Mating combinations and specialist genotypes

Queen gametes	Drone gametes			
	$P1_hP2_h$	$P1_hP2_l$	$P1_lP2_h$	$P1_lP2_l$
$P1_hP2_h$	$P1_hP1_hP2_hP2_h$	$P1_hP1_hP2_lP2_h$	$P1_lP1_hP2_hP2_h$	$P1_lP1_hP2_lP2_h$
$P1_hP2_l$	$P1_hP1_hP2_hP2_l$	$P1_hP1_hP2_lP2_l$	$P1_lP1_hP2_hP2_l$	$P1_lP1_hP2_lP2_l$
$P1_lP2_h$	$P1_hP1_lP2_hP2_h$	$P1_hP1_lP2_lP2_h$	$P1_lP1_lP2_hP2_h$	$P1_lP1_lP2_lP2_h$
$P1_lP2_l$	$P1_hP1_lP2_hP2_l$	$P1_hP1_lP2_lP2_l$	$P1_lP1_lP2_hP2_l$	$P1_lP1_lP2_lP2_l$

Note: The queen is a heterozygote for both loci and can produce four possible gametes with respect to these two loci. Male genotypes are haploid. Combinations for offspring from mating the heterozygous queen with all possible male genotypes are shown. Shaded boxes show the extreme specialist genotypes. There is no single mating that will produce progeny with both specialist genotypes.

genotype, and genetic variation will be maintained in the population. Multiple mating is also favored because it increases the variance in the genotypes and results in more of the specialist classes in a colony (Table 4.1).

4.4.1.2 Genetic Evidence for the Model   Genetic mapping of pollen- and nectar-foraging behavior has revealed major quantitative trait loci associated with the amount of pollen stored in the comb and pollen- and nectar-foraging behavior (Chapter 6). A quantitative trait locus (QTL) is a gene responsible for a quantitative trait. A quantitative trait is one that requires a measurement to show the phenotypic variation. For example, eye color in humans is a qualitative trait. It is classified as brown, blue, or green. However, height is quantitative; one must measure it to show the distribution of variation. We mapped QTLs for the amount of pollen stored in the comb (pollen hoarding) and found two major QTLs initially and a third in a subsequent map. These QTLs represented regions on a chromosome where more than one gene affecting the pollen-hoarding trait could reside. We subsequently constructed genomic maps for the individual traits of pollen and nectar foraging. They mapped to the same QTLs, demonstrating pleiotropy (when a gene has effects on more than one trait). Altogether we mapped three major

QTLs *(pln1, pln2, and pln3)* that affected the weight of nectar and pollen collected. Foraging loads of nectar and pollen are linked; they are not independent. We have collected thousands of foragers at entrances of hives, taken their loads, and weighed them. There is always a negative correlation (Figure 2.12). Bees with larger loads of pollen collect smaller loads of nectar, and vice versa. Therefore, these three QTLs are affecting the joint loading of pollen and nectar by foragers, as assumed in the model. Although the underlying genetic basis of observed variation in foraging for nectar and pollen is surely more complex than our model, the model does contain the essential elements: (1) there is genetic variation in populations for pollen and nectar collecting; (2) at least two genes are affecting foraging behavior; (3) both genes affect both pollen and nectar loads; and (4) there is no single optimum genotype. As a consequence, if the genetic specialists are better at foraging and this affects colony fitness, then polyandry should evolve.

4.4.1.3 Behavioral Evidence for the Model   Are genetic specialists better? The model assumes that this is the case and that polyandry is a mechanism to increase their numbers in a colony. In order to test for task performance of task specialists with different genotypes, we took advantage of the behavioral plasticity of honey bee foragers from strains of bees that were selected for the amount of surplus pollen they store, the pollen-hoarding strains (see Chapter 5 for complete description). Workers from the high-pollen-hoarding strain tend to specialize in pollen foraging, while those from the low-pollen-hoarding strain tend to be nectar specialists. However, there is a broad overlap in the foraging behavior of high- and low-strain bees. Some high-strain bees show a bias for nectar, and some low-strain bees bias their foraging toward pollen. Overall, the average foraging tendencies are dramatically different, but there is still much variation. Therefore, we were able to test individuals derived from each of these two strains as task specialists for both tasks.

The work described here was done by Gui Deng, a doctoral student in the laboratory of Keith Waddington. Waddington and I have collaborated over the years, and he spent two full-year sabbaticals in my lab doing research and taking up a new hobby—painting. Deng used bees from the high and low strains to test for foraging efficiency. Unfortunately, she dropped out of science after completing her doctoral degree and never published her results in scientific journals. However, they are contained in her doctoral dissertation at the University of Miami.

Queens of the high and low strains were instrumentally inseminated with sperm from a single drone each, derived from their own strain. The queens laid eggs in combs that were then placed together in a common nursery colony until larval development was complete. The nursery colony was a wild-type (not high- or low-strain) colony that was set up to raise the brood from eggs to pupae. The combs containing pupae were transferred to an incubator maintained at normal brood-nest temperature (about 35 degrees Celsius) where workers from each strain emerged into separate cages. Newly emerged workers of each strain were divided into two equal groups, marked with paint for strain identification, and placed in two host colonies. Each host colony received high- and low-strain workers. We call this our "common-garden experiment" (Figure 4.5). We wanted the marked "focal" bees to be in the minority in the hive because otherwise they might significantly affect the stimulus environment of the nest and, therefore, the behavior of the focal bees.

The two host colonies were placed in separate cages in the same indoor flight room. One host colony was offered only 50 percent sugar solution in its cage as a foraging resource, while the other was offered only freshly ground pollen. Bees foraged for these resources in the cage. When paint-marked workers began foraging, they were individually marked by gluing a numbered plastic disk onto the thorax. Bees were observed for three hours per day during their entire foraging lives. Nectar load weights were determined for each forager once each foraging day by placing a nectar feeder on an analytical balance and recording the net loss of weight from the feeder for that trip. Average load weights were determined for each forager of each strain, and the strain average was determined as the average of the individual averages. Pollen load weights were not determined because collecting pollen from

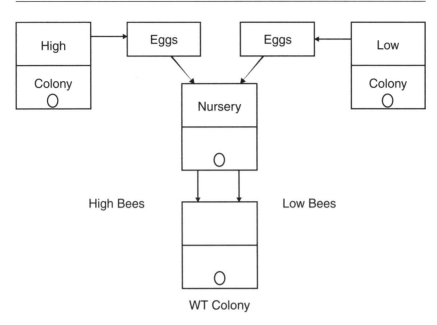

Figure 4.5. A common-garden experiment. High- and low-strain queens laid eggs in frames in their own hives. The frames were then removed and placed in an unrelated nursery colony where they were cared for at the same time by nurse bees that were unrelated to them. Combs were removed from the nursery colony before adult bees emerged and were placed in an incubator. When the adult bees emerged, they were marked with paint or individually tagged with numbered plastic disks and placed in a common wild-type colony to study their behavior.

dishes is a messy operation for bees. The bees scatter pollen outside the dish, which makes it impossible to get reliable weights without actually collecting the bees. This would disrupt their foraging behavior and might have long-term effects on foraging frequency.

As expected, some low-pollen-hoarding-strain workers collected pollen, and some high-pollen-hoarding-strain (high strain) workers collected nectar. However, the performance of foraging tasks varied with genotype. Low-pollen-strain (low strain) nectar specialists made more frequent foraging trips for nectar than did high-pollen-strain nectar specialists and carried larger nectar loads. High-pollen-hoarding-strain specialists made more pollen-foraging trips. Jennifer Fewell and I previously showed that pollen loads collected by high-strain pollen foragers are heavier than those of low-strain pollen foragers.

4.4.2 Rare-Genotype Specialists

The rare-genotype hypothesis of Stefan Fuchs and Robin Moritz proposes that task specialization has costs associated with it if too many individuals perform the same tasks. Colonies benefit by having a distribution of specialists in which specialists for any given task are rare. There could be merit in this hypothesis when one considers that some tasks probably do have negative effects on colony fitness if too many individuals perform them. In that case, we would expect that frequency-dependent selection could maintain genetic variation in the population. With frequency-dependent selection, an allele for a trait is favored by selection when it is rare but is detrimental when it is too common, similar to individual sex alleles in honey bees. One example might be the hygienic behavior discussed in Section 4.3.2. Hygienic behavior is a recessive trait (a bee must be homozygous for the genes that affect the behavior) and remains at relatively low frequencies in populations even though there is a clear advantage with respect to reducing parasite and pathogen loads. Why doesn't it increase in frequency? Perhaps there is a cost, as well as a benefit.

Chewing combs and removing wax from the nest are associated with hygienic behavior. Chewing combs can be important for removing combs that are contaminated or have reduced cell sizes because they are old and have built up many layers of cocoons. But chewing combs can also become pathological, from the perspective of a colony, if too many individuals engage in it. Kim Fondrk (personal observations) raised daughter queens from a select group of queens in our research apiary at the University of California–Davis and then instrumentally inseminated them with semen from their own brothers. He put them in hives in the apiary and let the colonies of inbred bees grow and develop. He noticed that some of the hives had piles of chewed wax in front of them. When he opened the hives, he noticed that the bees were chewing down all the wax on the brood combs, down to the midrib of the combs. He looked at the other colonies with piles of chewed wax in front and observed the same phenomenon. All the colonies displaying this behavior had sister queens who had been mated to drones derived from their mothers. He put new combs in, and they chewed them as well. They even chewed through cells that contained larvae and pupae and cast them out of the hive. Inbreeding, such as mating these queens to drones from their own colony, increases homozygosity at individual gene loci. Traits that are recessive and normally rarely seen, such as the genes for hygienic behavior, become homozygous, and the traits are expressed. Harry Laidlaw told me that he also had an inbred line at one time with the same behavior. He never saw the trait except when he inbred. The same is true in our experience.

4.4.3 Other Evidence for the Genetic-Specialist Hypothesis

Several studies have attempted to do direct tests comparing colonies with multiply inseminated queens with those with queens that are singly inseminated. Most of the studies have controlled the sources of queens and drones used for insemination so that the single- and multiple-insemination groups have been balanced with respect to genotypes (Figure 4.6). For example, drones are taken from several lines, or queen sources, and are used for single insemination of sister queens. Then other sister queens are inseminated with sperm derived from equal contributions of drones from combined pairs of sources or all sources pooled. Colonies headed by queens that were singly inseminated with sperm from males from different lines or queen sources are compared. Observed trait differences among drone sources suggest genetic variation for the traits. Then the queens with pooled sperm from multiple sources (the more genetically diverse groups) are compared with the averages for the individual sources from which the drones were combined for insemination. Deviation from the expected average values of the individual sources suggests an effect of genetic variation, a consequence of the interactions of the genotypes. Experiments of this type have failed to give clear demonstrations of effects of genetic diversity on colony traits. The reasons for failure could be the limited number of drone

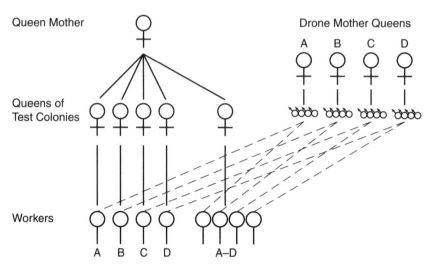

Figure 4.6. A cross designed to test the effects of genetic diversity.

source combinations tested (two to six), resulting in much less potential genetic variation compared with open-mated queens, small sample sizes, or both.

The best demonstrations of the effects of multiple mating are those of Heather Mattila. She and her collaborators set up colonies with queens that had been inseminated with just one male or with a pooled and mixed combination of 15 males. All queens received the same amount of semen during instrumental insemination. Drones were drawn at random from a pool of about 1,000 drones from 11 source colonies. Altogether, 12 genotypically diverse and 9 genotypically less diverse (single-male) colonies were tested. Colonies were set up as swarms and were allowed to build up and make it through the winter. They were fed antibiotics to control for any effects of genotypic diversity on parasite and pathogen loads in the colonies. Colonies with poor brood viability were removed, a control for the production of diploid males. The results were clear: the more genotypically diverse colonies built up their colonies faster (combs, food, and bees), had more foraging activity, stored more food, had more adult workers, survived the winter better, and raised more reproductive males. The success of this study was apparently due to the much larger number of males contributing to the inseminations of the queens of the high-diversity group, generating much more variance than earlier studies, and probably the more severe and punctuated seasons in Ithaca, New York.

Colonies that are more genotypically diverse have more dance communication, perhaps one reason they are more successful in general. Pollen-foraging genetic specialists are more effective recruiters for pollen (Section 2.5.7.4). In addition, genetic specialists are more likely to find the resources they are genetically tuned to when they act as scouts, and they spend more time attending dances for their genetic bias (Section 5.3.2).

4.4.4 Genotypic Variation Facilitates Graded Responses to Environmental Change

Another form of the genetic-variation hypothesis for polyandry proposes that colonies that are more genotypically diverse are less likely to fail because they are more average for their traits. Colonies with less variability may do better for some traits but fail at others, reducing their chances to survive and reproduce. But colonies with more variation do everything at or near the population average, which represents the historical optimum. We tested 43 colonies with super-sister queens. Queens were instrumentally inseminated either with four males from one of four source queens (low diversity) or with one male from each of the four sources (high diversity; Figure 4.6). Colonies were allowed to grow in worker population, and then all their food, comb, and brood were taken away from them, and they were forced to start over. They were observed for 21 days, the time it takes for bees to develop from egg to adult, and were tested for 19 colony traits. The experiment was terminated after 21 days because that was the time at which new bees could enter the population. So this test represented the efforts of single cohorts of bees over 21 days.

Significant genetic variation was found for seven traits. Effects of genotypic diversity were tested by comparing the means of the diverse colonies with the combined mean of the single-drone-inseminated colonies of all four sources. Only two traits varied, not more than expected due to chance with so many comparisons. The distributions of the ranks of the means (1–5 corresponding to the five treatment groups, four single sources and one combined source) for the 19 traits revealed that the high-diversity colonies had more intermediate ranks for traits. Genetic variation made the colonies more average.

4.5 A Pluralistic View of the Evolution of Polyandry

It is unlikely that a single model can explain the evolution of polyandry in honey bees, much less all the Hymenoptera. Instead, it is likely that there are multiple explanations. The potential effects of diploid drone production on mating-behavior evolution have been clearly demonstrated. But the mechanism of sex determination cannot explain the high levels of recombination in bees because the production of diploid males is a consequence of a single gene, and, therefore, recombination has no effect. So, there must be additional benefits derived from increased genetic variation. The other hypotheses are more consistent with high levels of recombination as long as the traits involved are polygenic. Selection to avoid parasites and pathogens probably has played a role but probably is not a gene-for-gene relationship of host and parasite. It is more likely that broad-based, polygenic resistance is involved; however, the empirical results are still equivocal. The generally higher performance of more genetically diverse colonies has been clearly demonstrated. We know that the general effects are in addition to any effects of polyandry on the consequences of producing diploid males and are distinct from the effects on parasites and pathogens. We still do not know to what degree there is a need for rare genetic specialists, a need for a broader distribution of genetically programmed specialists, or a need for colonies to be near average states for as many traits as possible in order to avoid failure.

Suggested Reading

Bambrick, J. F. 1964. Resistance to American foulbrood in honey bees. V. Comparative pathogenesis in resistant and susceptible larvae. *J. Insect Pathol.* 6:284–304.

Bambrick, J. F. 1967. Resistance to American foulbrood in honey bees. VI. Spore germination in larvae of different ages. *J. Invert. Pathol.* 9:30–34.

Bambrick, J. F., and Rothenbuhler, W. C. 1961. Resistance to American foulbrood in honey bees. IV. The relationship between larval age at inoculation and mortality in a resistant and in a susceptible line. *J. Insect Pathol.* 3:381–390.

Beye, M., Hasselmann, M., Fondrk, M. K., Page, R. E., et al. 2003. The gene *csd* is the primary signal for sexual development in the honeybee and encodes an SR-type protein. *Cell* 114:419–429.

Beye, M., Hunt, G. J., Page, R. E., Fondrk, M. K., et al. 1999. Unusually high recombination rate detected in the sex locus region of the honey bee *(Apis mellifera). Genetics* 153:1701–1708.

Cale, G. H., and Rothenbuhler, W. C. 1975. Genetics and breeding of the honey bee. In *The Hive and the Honey Bee.* Hamilton, IL: Dadant and Sons, pp. 157–184.

Crozier, R. H., and Fjerdingstad, E. J. 2001. Polyandry in social hymenoptera—Disunity in diversity? *Ann. Zool. Fennici* 38:267–285.

Crozier, R. H., and Page, R. E. 1985. On being the right size: Male contributions and multiple mating in social Hymenoptera. *Behav. Ecol. Sociobiol.* 18:105–115.

Deng, G. 1996. Foraging performance of honey bees *(Apis mellifera).* PhD diss., University of Miami.

Fuchs, S., and Moritz, R. F. A. 1998. Evolution of extreme polyandry in the honeybee Apis mellifera L. *Behav. Ecol. Sociobiol.* 9:269–275.

Gempe, T., Hasselmann, M., Schiøtt, M., Hause, G., et al. 2009. Sex determination in honeybees: Two separate mechanisms induce and maintain the female pathway. *PLoS Biol* 7:e1000222. doi:10.1371/journal.pbio.1000222.

Hamilton, W. D. 1972. Altruism and related phenomena, mainly in social insects. *Ann. Rev. Ecol. Syst.* 3:193–232.

Hasselmann, M., Fondrk, M. K., Page, R. E., and Beye, M. 2001. Fine scale mapping in the sex locus region of the honey bee *(Apis mellifera). Insect Mol. Biol.* 10:605–608.

Hunt, G. J., and Page, R. E. 1994. Linkage analysis of sex determination in the honey bee *(Apis mellifera). Mol. Gen. Genet.* 244:512–518.

Hunt, G. J., and Page, R. E. 1995. A linkage map of the honey bee, *Apis mellifera,* based on RAPD markers. *Genetics* 139:1371–1382.

Kraus, B., and Page, R. E. 1998. Parasites, pathogens, and polyandry in social insects. *Am. Nat.* 151:383–391.

Laidlaw, H. H., and Page, R. E. 1984. Polyandry in honey bees: Sperm utilization and intra-colony genetic relationships. *Genetics* 108:985–997.

Laidlaw, H. H., and Page, R. E. 1997. *Queen Rearing and Bee Breeding.* Cheshire, CT: Wicwas Press.

Lewis, L. F., and Rothenbuhler, W. C. 1961. Resistance to American foulbrood in honey bees. III. Differential survival of the two kinds of larvae from two-drone matings. *J. Insect Pathol.* 3:197–215.

Mattila, H. R., Burke, K. M., and Seeley, T. D. 2008. Genetic diversity within honeybee colonies increases signal production by waggle-dancing foragers. *Proc. R. Soc. B* 2008 275:809–816.

Mattila, H. R., and Seeley, T. D. 2007. Genetic diversity in honey bee colonies enhances productivity and fitness. *Science* 317:362–364.

Oldroyd, B. P., and Fewell, J. H. 2007. Genetic diversity promotes homeostasis in insect colonies. *Trends Ecol. Evol.* 22:408–413.

Page, R. E. 1980. The evolution of multiple mating behavior of honey bee queens. *Genetics* 96:263–273.

Page, R. E., Gadau, J., and Beye, M. 2002. The emergence of hymenopteran genetics. *Genetics* 160:375–379.

Page, R. E., and Laidlaw, H. H. 1988. Full sisters and super sisters: A terminological paradigm. *Anim. Behav.* 36:944–945.

Page, R. E., and Metcalf, R. A. 1982. Multiple mating, sperm utilization, and social evolution. *Am. Nat.* 119:263–281.

Page, R. E., and Metcalf, R. A. 1984. A population investment sex ratio for the honey bee. *Am. Nat.* 124:680–702.

Page, R. E., Robinson, G. E., Fondrk, M. K., and Nasr, M. E. 1995. Effects of worker genotypic diversity on honey bee colony development and behavior (*Apis mellifera* L.). *Behav. Ecol. Sociobiol.* 36:387–396.

Palmer, K. A., and Oldroyd, B. J. 2000. Evolution of multiple mating in the genus *Apis. Apidologie* 31:235–248.

Rinderer, T. E., and Rothenbuhler, W. C. 1969. Resistance to American foulbrood in honey bees. X. Comparative mortality of queen, worker, and drone larvae. *J. Invert. Pathol.* 13:81–86.

Rothenbuhler, W. C. 1964. Behavior genetics of nest cleaning in honey bees. IV. Response of F1 and back-cross generations to disease-killed brood. *Am. Zool.* 4:111–123.

Rothenbuhler, W. C. 1967. American foulbrood and bee biology. In *Twenty-First International Beekeeping Congress, College Park, Maryland,* ed. V. Harnaj. Bucharest: Apimondia, pp. 179–188.

Rothenbuhler, W. C., and Thompson, V. C. 1956. Resistance to American foulbrood in honey bees. I. Differential survival of larvae of different genetic lines. *J. Econ. Entomol.* 49:470–475.

Seeley, T. D., and Tarpy, D. R. 2007. Queen promiscuity lowers disease within honeybee colonies. *Proc. R. Soc. B* 274:67–72.

Sherman, P. W., Seeley, T. D., and Reeve, H. K. 1988. Parasites, pathogens, and polyandry in social Hymenoptera. *Am. Nat.* 131:602–610.

Sherman, P, W., Seeley, T. D., and Reeve, H. K. 1998. Parasites, pathogens, and polyandry in honey bees. *Am. Nat.* 151:392–396.

Sutter, G. R., Rothenbuhler, W. C., and Raun, E. S. 1968. Resistance to American foulbrood in honey bees. VII. Growth of resistant and susceptible larvae. *J. Invert. Pathol.* 12:25–28.

Tarpy, D. R. 2003. Genetic diversity within honeybee colonies prevents severe infections and promotes colony growth. *Proc. R. Soc. B* 270:99–103.

Tarpy, D. R., and Page, R. E. 2000. No behavioral control over mating frequency in queen honey bees (*Apis mellifera* L.): Implications for the evolution of extreme polyandry. *Am. Nat.* 155:820–827.

Tarpy, D. R., and Page, R. E. 2002. Sex determination and the evolution of polyandry in honey bees *(Apis mellifera). Behav. Ecol. Sociobiol.* 52:143–150.

Thompson, V. C., and Rothenbuhler, W. C. 1957. Resistance to American foulbrood in honey bees. II. Differential protection of larvae by adults of different genetic lines. *J. Econ. Entomol.* 50:731–737.

Tooby, J. 1982. Pathogens, polymorphism, and the evolution of sex. *J. Theor. Biol.* 97:557–576.

Trivers, R. L., and Hare, H. H. 1976. Haplodiploidy and the evolution of social insects. *Science* 191:249–263.

Van Valen, L. 1973. A new evolutionary law. *Evol. Theory* 1:1–30.

Woyke, J. 1964. Causes of repeated mating flights by queen honeybees. *J. Apic. Res.* 3:17–23.

— 5 —

The Phenotypic Architecture
of Pollen Hoarding

In Chapter 2, I introduced the stone-soup model of how social organization with division of labor arises from the simple mechanism of individuals responding to stimuli in their environment, changing the stimuli through their actions, and thereby affecting the behavior of nestmates. I called this the "spirit of the hive," an answer to Maurice Maeterlinck's puzzle. In Chapter 3, I introduced genetic variation into the pot, building some additional complexity into the stone soup and changing the flavor. Chapter 4 showed how the polyandrous mating system that gives rise to increased genetic variation could evolve and links the mating behavior of the queen with the genetically diverse social environment that she produces through her egg gametes and the sperm of her mates. In this chapter and Chapter 6, I will build a complex network of genes and phenotypes, mapping behavioral, anatomical, physiological, developmental, and genetic mechanisms onto social foraging behavior.

5.1 Levels of Biological Organization

How is complex social behavior organized with respect to the different levels of biological organization from which societies are built: genetic regulatory systems, proteins, cell signaling, hormonal regulatory networks, developmental cascades, individual anatomy, morphology, and behavior, and colony-level traits derived from the interactions of tens of thousands of individuals? This is truly a mind-boggling, complex problem. There is no central control of the activities of individual workers: they have limited global information about the state of the nest and the activities of others and behave by responding to local stimuli. From these different organizational levels emerges self-organized collective behavior. To better understand the mechanisms operating at the different organizational levels, I use a model social phenotype—the amount of surplus pollen stored by bees.

My approach to this question has been to combine selective breeding with mechanistic studies of behavior and physiology. I view selective breeding as analogous to natural selection, except that with selective breeding I, rather than the environment, am the agent of selection. Darwin used the analogy of selective breeding repeatedly in *On the Origin of Species* when he drew on the results of breeders of domestic pigeons to show how selection can result in dramatic changes in phenotypes. The same processes take place with natural selection and artificial selection at the level of changes in the frequencies of heritable phenotypes from generation to generation. Allelic variants of genes affecting the traits I select increase or decrease in frequency in my breeding populations just as they do in natural populations. Response to my selection is attributed to genes that are variable and affect my traits of interest, and genes that have the biggest effects on the phenotype increase or decrease in frequency more rapidly than genes with smaller effects. The underlying changes in anatomy, physiology, and biochemistry are linked to the changes in gene frequencies, again, just as in natural populations. The differences between natural selection on natural populations and my artificial selection on closed populations are the size of the breeding population and the strength of the selection. My populations are small, and I apply strong selection, while natural populations are much larger, and selection for given phenotypic traits is not expected to be as severe. As a consequence, with my selection program phenotypic change is faster, genetic variation is exhausted faster, and phenotypic changes are dominated by genes with large effects. Nonetheless, my program provides a tool to study the effects of selection across levels of biological organization.

5.2 Selective Breeding for Pollen Hoarding

In 1986, I took my first faculty position at The Ohio State University. Walter Rothenbuhler had just retired, and I was fortunate to be hired as his successor, an honor for an aspiring behavioral geneticist. Rothenbuhler was one of the pioneers of behavioral genetics and had taken Rick Hellmich as a graduate student to work on an artificial-selection project to study the heritability of pollen hoarding, the amount of surplus pollen stored by colonies. They produced two strains of bees that differed in the amount of pollen they hoarded. They estimated the heritability (h^2) of the trait to be 49.5 (see Section 5.2.1 for an explanation of heritability). This was a very high heritability for a behavioral trait. The main purpose of the project was to breed bees to store more pollen in order to be more active pollinators and have better colony nutrition.

When I arrived, the strains were being maintained by Nick Calderone, a former master's student of Rothenbuhler's, and Kim Fondrk, Rothenbuhler's technician. Calderone wanted to get a doctorate and continue working with the stocks. I was interested in looking at the effects of selection for a colony-level phenotype on individual behavior. Calderone took on a doctoral project under my direction. The results of his important studies are discussed in Chapter 3.

In 1989, I was offered a faculty position at the University of California–Davis, where I had done my doctoral work. I accepted the position and moved back. Kim Fondrk moved as well and continued to be my technician for more than 25 years. I was truly fortunate to have him as a technician and research partner. Before the move, we discussed the fate of the pollen-hoarding strains of Hellmich and Rothenbuhler. Should we take them with us or start over? We decided to initiate a new selection study in California, using the same selection criteria, but this time to focus on, and document, the effects of the selection on the different levels of biological organization from social phenotype to genes.

5.2.1 Heritability

I think that heritability needs more explanation because it is an important concept in breeding. Heritability, often shown as h^2, is the proportion of the total variance observed in a sample of phenotypes that can be explained by differences in genotype. Variance is a measure of the breadth of a distribution of phenotypes around the average. For example the set of numbers {4, 4, 5, 6, 6} has a mean (average) value of 5. So does the set {3, 4, 5, 6, 7}. But which one has more variance? The second set, of course.

Imagine the population of Europe. When you travel from country to country, you notice that people look somewhat different. This is still true today even with modern globalization. When I took my first trip to the Netherlands, I was struck by how tall people were. Walking through the train station at 1.75 meters, I felt short. In fact, men and women in the Netherlands constitute the tallest population on earth. Men on average are 1.84 meters tall (Wikipedia); women on average are 1.70 meters tall. If you randomly sampled the heights of men across the Netherlands, you would have a distribution with a broad range. Some of that range would be due to developmental events influenced by diet, disease, or other factors, but some of it would be due to the genes they inherited from their parents. There are statistical methods that allow you to estimate the proportion of the variance in height that is due to the height of the parents, the heritable portion, if you are able to sample the parents as well as their offspring. The heritable portion can be further subdivided into additive and nonadditive parts. The additive portion is called heritability in the narrow sense and is the additive effect of each allele at each quantitative gene; the nonadditive effect includes all the interactions of the genes and the interactions of the genes with the environment.

You could ask what proportion of the variance in height is due to being a male or female. You would find that gender is a significant contributor to the distribution of heights in the total, combined population. Or you could take a random sample of men from New York City who came from immigrant French and Dutch families. French men have an average height of 1.74 meters. Then ask the question: "What proportion of the total variance in the sample is a result of the men being from French or Dutch parents?" In that case, both the genetic complement of the populations and the shared environmental effects of growing up in New York City would be responsible. This is not heritability in the sense in which it is used in quantitative genetics, but it is a measure of the degree of genetic determination. This is how we compared bees from the high- and low-pollen-hoarding strains.

5.2.2 The Social Phenotype

We selected for a single trait, the amount of pollen stored in the comb—pollen hoarding. This is a colony-level, social trait that is a consequence of the interactions of tens of thousands of individuals in the nest. Foragers collect pollen and store it in the comb. Nurse bees consume the pollen and convert the pollen proteins into glandular secretions that they feed to larvae. Assume that a queen lays about 1,500 eggs per day during spring and summer; that a worker lives for 6 weeks, for 1 week of which she is a nurse bee and 3 weeks a forager; and that about one-third of foragers collect pollen. Larvae are fed for about 6 days. With these assumptions, a colony will consist of about 10,000 pollen foragers, 10,000 nurse bees, and 9,000 larvae consuming pollen being processed through the nurse bees. The pollen foragers are engaged in assessing the amount of stored pollen and the number of young larvae being fed, foraging, and performing recruitment dances to engage other foragers in pollen collecting. Thus the amount of stored pollen is the result of a highly interactive social, regulatory network (Figure 5.1).

Pollen is stored near the brood. The amount of stored pollen is regulated (see Section 1.2 and Chapter 9), which makes it an ideal phenotype for selection (Figure 5.2). The regulation of stored pollen results in high repeatability of measurements and consequently higher heritability because there is less measurement error and environmental "noise" associated with the phenotype. The amount of stored honey is not regulated. Bees will continue to store surplus honey as long as there are empty cells in combs and nectar is available; therefore, the heritability is much lower and usually is not statistically significant. We estimated the amount of pollen stored in the combs of colonies by examining individual frames of combs containing pollen. We placed a wire grid over the combs and counted the number of grid squares covering the areas of stored pollen, from which we estimated the total area of stored pollen in the hive (Figure 5.2).

Pollen Regulation

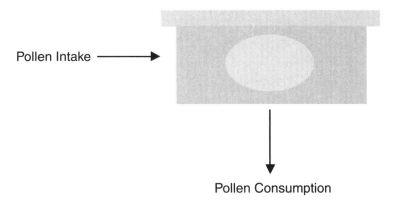

Pollen Intake ⟶

Pollen Consumption

Figure 5.1. Honey bee colonies regulate pollen. It is stored close to the brood where nurse bees access the stored pollen, consume it, and convert it into proteinaceous glandular secretions fed to developing larvae. Pollen foragers assess the quantity of stored pollen and the pollen need, indicated by pheromones produced by the larvae, and bias their foraging trips accordingly for pollen (protein) and nectar (carbohydrate). Successful foragers perform recruitment dances based on their perception of the quality of the pollen and recruit additional pollen foragers.

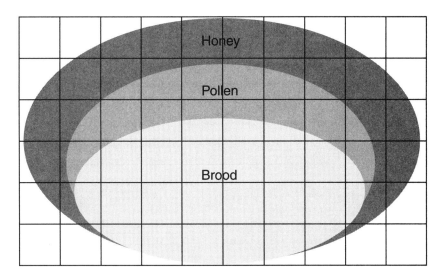

Figure 5.2. The organizational pattern of stored pollen. Measurements were made of the area of stored pollen on combs such as this by placing a grid over the comb and counting the number of grid squares that had pollen under them.

5.2.3 Selection for Pollen Hoarding

We initiated the Davis pollen-hoarding selection study in the spring of 1990. We examined about 400 commercial hives located in almond orchards near Davis, California (Figure 5.3). We selected 127 colonies with approximately equal worker populations and measured the amount of stored pollen. The 10 colonies with the most stored pollen and the 10 colonies with the least were selected to initiate the high and low strains, respectively. Virgin queens and drones were raised from each of the colonies, and queens were mated to single drones by instrumental insemination (Figure 5.4) to establish five sublines of highs and lows. Each generation, new virgin queens and drones were raised from the superior-performing queen of each subline, and matings were performed between lines according to a circular breeding system (Figure 5.5).

We were interested in the effects of the selection on individual behavior. The studies of Nick Calderone showed that the high-strain bees of

Figure 5.3. Kim Fondrk inspecting hives in an almond orchard near Davis, California. Photo by Kathy Keatley Garvey.

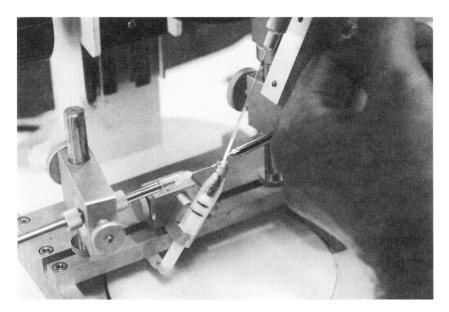

Figure 5.4. Instrumental insemination of a queen honey bee. The queen is positioned head down, ventral side up. The hook on the right moves the sting out of the way to make room for the syringe, full of semen, to be placed in the vaginal orifice. Semen is injected into the median oviduct. Photo by Jacob Sahertian.

Hellmich and Rothenbuhler were biased toward collecting pollen and foraged one day earlier in life than the low strain. The bias in pollen foraging was an obvious and expected consequence. Other hypothesized mechanisms were the following: (H1) high-strain colonies have more larvae and hence more brood pheromone to stimulate pollen foraging; (H2) high-strain colonies produce fewer larvae and thereby consume less pollen; (H3) high-strain individual larvae produce more pollen-foraging cues; (H4) high-strain queens produce cues that stimulate more pollen foraging; (H5) high-strain colonies have more foragers; and (H6) high-strain colonies have a higher proportion of pollen foragers.

The response to selection was very rapid. After the first generation of selection, the amounts of stored pollen of the high and low strains were significantly different, demonstrating that a significant amount of the variation observed in the commercial hives with naturally mated queens was due to genetic variation among colonies. By generation 3,

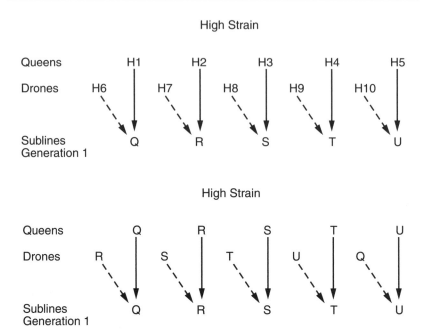

Figure 5.5. Breeding scheme for the high strain. The upper panel shows how the foundation generations were established for the high strain (the selection process for the low strain is identical). The 10 colonies with highest amounts of stored pollen were chosen as parents and designated H1 to H10. Queens were raised from H1 to H5, while drones were raised from H6 to H10. Drones from H6 were used to inseminate (by instrumental insemination) queens from H1, and so on. Queens were inseminated with the semen of a single drone each. The lower panel shows the pairings of sublines in generation 2. Each generation the drone source rotated to the next subline; for example, in generation 3, queens from the Q-subline were inseminated with drones from the S-subline. Reprinted with kind permission from Springer Science+Business Media: *Behavioral Ecology and Sociobiology*, "The effects of colony-level selection on the social organization of honey bee (*Apis mellifera* L.) colonies: Colony-level components of pollen hoarding," 36(2), 1995, 134–144, Page RE, Fondrk MK, Fig. 1.

the high-strain bees stored 6 times more pollen than bees from the low strain. After 32 generations, the high-strain bees stored 12 times more pollen. In generation 4, we tested the high- and low-strain colonies against the commercial bees from which we established our strains and found that the commercial bees were intermediate in phenotype. This showed that both the high and low strains responded to selection: the response was bidirectional. Selection for increased pollen hoarding did not affect the amount of stored honey, but by generation 3, the brood area of high-strain colonies was significantly less (22 percent) than that of colonies of the low strain. The reduced brood area was a consequence of the limited space in the single-box test hives (a standard single box has a volume of about 40 liters and holds 10 frames containing comb). Bees store pollen close to the brood. The high-strain bees store so much that it restricts the space available for egg laying and brood rearing. In generation 4, we tested highs and lows in two-chamber hives (hives composed of two boxes; most commercial hives have two or more) and found no differences in the amount of brood reared, but still large differences in stored pollen. Therefore, we rejected H1 and H2. In generation 33, we again looked at the amount of brood and honey and the worker populations of high- and low-strain colonies. Total brood was reduced by 23 percent, as expected because of the constraints of the stored pollen on brood production; there were still no differences in honey; and the worker population of high-strain colonies were slightly, but statistically significantly lower (18 percent), probably because of reduced brood production and the shorter life span of high-strain workers (Section 5.3.1).

For generation 3, we set up colonies with wild-type bees and high- or low-strain queens. We evaluated colonies 30 days later, when the first high- and low-strain workers were emerging. We evaluated them three more times as they developed over about 7 weeks. During this time, the wild-type workers were dying (bees that live long enough to become foragers live about 6 weeks at this time of year), and more high- and low-strain workers were produced and took over the roles of foragers (it takes 21 days for a worker to develop from egg to adult and then another 2 to 3 weeks to become a forager). High- and low-strain test hives did not differ in the amount of stored pollen until the third evaluation, approximately 60 days after the colonies were established. This is the amount of time required for the high- and low-strain bees derived from the queens to take over as the forager population in the hive. During this period, however, the wild-type foragers were exposed to high- and low-strain queens and brood, but the strains did not diverge in stored pollen until the forager populations changed to the high- and low-strain bees (Figure 5.6). These results clearly showed that queen and larval cues were not responsible for differences in stored pollen between the strains.

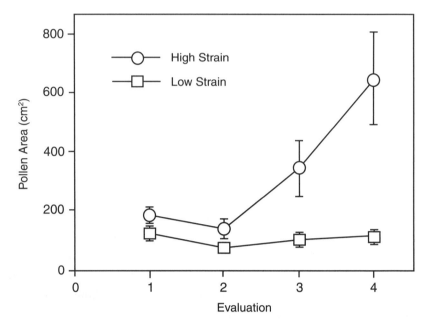

Figure 5.6. Area of pollen (cm²) for high- and low-strain colonies for the four evaluations of generation 3. Means and 95 percent confidence intervals are presented for 29, 30, 30, and 30 low-strain colonies and 23, 27, 27, and 27 high-strain colonies for evaluations 1–4, respectively. Reprinted with kind permission from Springer Science+Business Media: *Behavioral Ecology and Sociobiology,* "The effects of colony-level selection on the social organization of honey bee (*Apis mellifera* L.) colonies: Colony-level components of pollen hoarding," 36(2), 1995, 134–144, Page RE, Fondrk MK, Fig. 2.

Additional support comes from cross-fostering experiments. High- and low-strain bees were raised together in high- and low-strain hives with high- and low-strain queens and brood. High-strain foragers collected more pollen in the low-strain hive, and low-strain foragers collected less pollen in the high-strain hive. This result is the opposite of what we expected on the basis of H3 and H4; therefore, those hypotheses were rejected. Although high- and low-strain colonies differed in the proportions of their foragers collecting loads of pollen, they did not differ in total number of foragers and in generation 33 had fewer, rather than more, total bees because of the reduced amount of brood reared, so we rejected H5 (Table 5.1).

Table 5.1 Colony-level components of pollen hoarding

Component	Effect	Generation tested	Selected	Selectable
Queen cues	Increased pollen-foraging stimulus	3	No	?
Brood cues	Increased pollen-foraging stimulus	3	No	?
Brood quantity	Increased pollen-foraging stimulus, increased pollen consumption, and increased worker population	3, 33	Yes[a]	Yes
Worker population	Potential increase in total forager population	3, 33	Yes[b]	Yes
Total foragers	Increase in pollen foragers	3	No	Yes[c]
Number of pollen foragers	Increase in pollen intake	3, 5, others	Yes	Yes
Proportion of pollen foragers[d]	Increase in pollen intake	3, 5, others	Yes	Yes

Note: Shown here are potential colony-level traits that could increase the amount of pollen stored in colonies. Some of these traits did change as a consequence of selection for pollen hoarding, but others did not. In the "Selected" column, a trait was considered selected if there were differences between the high and low strains. Traits were selectable if differences were observed between sublines within the strains but not between the strains, for instance, between Q and R in Figure 5.5.

a. This effect is based on the constraint placed on the brood-nest size by the storage of surplus pollen.

b. This effect is based on the constraint placed on rearing of brood by the storage of surplus pollen, and the shorter life span of high-strain bees; however, it did not increase the number of foragers.

c. Significant variation was found between sublines of highs and lows, but there were no mean differences between the strains. This showed that genetic variation existed but had not been selected.

d. Proportion is calculated as (number of pollen foragers)/(total foragers).

Colonies composed of workers that were hybrids of the high and low strains performed like low-strain colonies for pollen hoarding, demonstrating directional dominance for the low-pollen-hoarding trait. However, they stored significantly more honey than either strain.

5.3 Individual Behavior

In the following sections, I will detail many of the experiments and results that have resulted in a growing understanding of the phenotypic architecture of foraging behavior. The effects of selection on pollen-hoarding behavior were as we expected on the basis of the study of Rick Hellmich and Walter Rothenbuhler. To study individual behavior of workers, we conducted common-garden experiments (Figure 4.5). Bees were collected as foragers. We tried to collect bees during their first day of foraging when we studied foraging-initiation age (age of first foraging). However, for studies of foraging behavior, we usually let a group of foragers build up for each of the genetic groups we wanted to compare. We then collected large samples of foragers over a few contiguous days. Collecting bees when they initiate foraging is less reliable for resource-foraging behavior studies because collections are often made over several weeks during which the weather and available flowers have large effects on the loads collected (that is, there is more environmental variance). Bees that initiate foraging later in life experience different foraging environments than those that initiate foraging earlier.

5.3.1 Age of Initiation of Foraging and Life Span

High-strain bees initiate foraging up to 12 days earlier in life. This result was unexpected but consistent with the results of the selection by Hellmich and Rothenbuhler. The life span of a bee can be divided into two parts: preforaging and foraging. The onset of foraging marks a major transition for bees with major consequences for mortality. Bees that initiate foraging later in life live longer, overall, but have shorter foraging spans. Every day of life before the initiation of foraging "costs" one-third of a day of foraging life. As a consequence, low-strain bees live longer than high-strain bees because they initiate foraging later.

5.3.2 Foraging Bias

Nectar and pollen loads are not independent. High-strain foragers collect relatively larger loads of pollen and smaller loads of nectar, so pollen is a relatively higher proportion of their total loads. High- and low-strain bees do not differ in the floral types of pollen they collect. High-strain bees are more likely to collect water and under poorer foraging conditions are more likely to return with loads of nectar that are more dilute, while low-strain bees are more likely to return empty.

As we found for the pollen-hoarding trait, pollen-foraging bias shows directional dominance for low pollen bias, with the low strain more similar to the wild-type bees. Relatively more high-strain bees collect only pollen, and more low-strain bees collect only nectar, but many collect both (Figure 5.7). Hybrid workers derived from crossing high- and low-strain queens and drones show an even stronger bias for nectar than the low-strain workers, corresponding to the greater amount of stored honey found for hybrid colonies. High- and low-strain bees respond to changes in the nest stimulus environment. Both collect more pollen and forage at younger ages under the high-pollen-foraging stimulus of high quantities of young larvae and low stored pollen. However, high-strain bees are more sensitive to the pollen-foraging stimuli and more readily change their foraging behavior in response to changes in brood and stored-pollen stimuli. High-strain bees are less likely to return empty from a foraging trip. Wild-type bees are intermediate in foraging behavior.

Potential foragers (recruits) may be biased to attend recruitment dances and get information from pollen or nectar foragers. If this is the case, it could explain some of the tendency for high-strain bees to forage for pollen and low-strain bees to forage for nectar. Claudia Dreller marked a large number of high- and low-strain bees and put them in an observation hive where she could observe their behavior

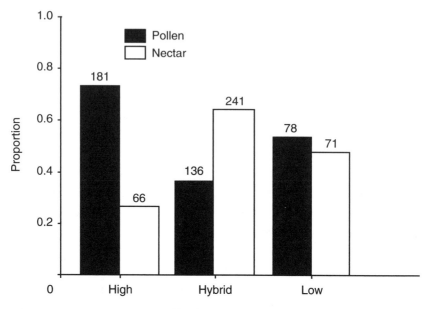

Figure 5.7. Foraging activity of high- and low-strain workers and hybrids from a common-garden experiment. Results are from four test colonies that were maintained in cages with feeders containing sugar solution (nectar) or pollen. Numbers on the top of the bars are the number of foragers observed. Reprinted from *Animal Behaviour,* 50(6), Page et al., "Genetic determinants of honey bee foraging behaviour," 1617–1625, Fig. 2 (1995), with permission from Elsevier.

(unpublished data). She found that high- and low-strain bees attended dances at random—they simply bumped into them, as found by Tom Seeley and William Towne. However, high-strain bees attended more cycles of waggle dances for pollen, and low-strain bees attended more for nectar. In other words, each strain showed a bias for what it paid attention to. Dreller also found that high-strain scout bees (bees that were not recruited to forage by attending a recruitment dance) were more likely to return with a load of pollen after making their first flight, while low-strain scouts were more likely to return with nectar. The collective biases for scouting and dance attendance should lead to significant biases in the foraging efforts of high- and low-strain colonies.

5.4 Sensory-Response Systems

Throughout our studies of the effect of selection on the high- and low-pollen-hoarding strains, we have conducted parallel studies using wild-type bees. Studies of wild-type bees that were conducted in the laboratory of Jochen Erber of the Technical University of Berlin motivated us to look at the same questions in the highs and lows. Studies of the high- and low-strain bees in my lab motivated him to look for similar phenomena in wild-type bees. The end result has been the ability to generalize our findings outside the context of our selectively bred bees.

5.4.1 Sensitivity to Sucrose

Our studies of the sensory-response systems in high- and low-strain bees were motivated by experiments initially conducted in Erber's lab when I was on sabbatical leave in 1996 (Section 2.5.2). We tested returning pollen and nectar foragers from a colony of Carniolan bees *(Apis mellifera carnica)* for their proboscis extension response to water and sucrose solutions and found that they differed significantly (Figure 2.7). We thought that they might differ fundamentally, perhaps at the level of the central nervous system, for their sensory perception and response, but we did not know how to uncouple the experiential effects of pollen and nectar foraging after they had already initiated foraging, so we had no way to know what they were "predisposed" to forage for before they actually returned with a load. The behavior itself might be tuning sucrose responsiveness. The high- and low-strain bees gave us a tool to uncouple genetically determined biases from foraging experience because we had groups of bees that we knew had some level of genetic predetermination to bias their foraging (Figure 5.8). Therefore, we tested them.

5.4.1.1 High- and Low-Strain Bees and Sucrose Sensitivity   High-strain bees are more responsive to water and more sensitive to sucrose than are low-strain bees, which explains why high-strain foragers are more likely to collect water and more dilute nectar. We tested high- and

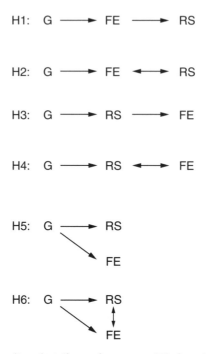

Figure 5.8. Disentangling the effects of genotype (G), foraging experience (FE), and response to sugar (RS). Given that there are genotypic differences in foraging behavior and in responsiveness to sugar, there are six different cause-and-effect hypotheses (H1 to H6) that can explain the relationship between foraging behavior and RS. Directions of the arrows show the hypothesized direction of causality.

low-strain pollen and nonpollen (presumably nectar) foragers. They were collected as they returned to the hive from a foraging trip and were tested for their response to a single stimulus of 30 percent sucrose solution by touching a droplet to their antennae. High-strain foragers responded by extending the proboscis significantly more often than did low-strain foragers. High-strain pollen foragers responded 48 percent of the time, low-strain pollen foragers 25 percent, high-strain nonpollen foragers 30 percent, and low-strain nonpollen foragers 17 percent. High- and low-strain nonpollen foragers were significantly different, but the sample size for low-strain pollen foragers was too low for statistical analyses, with just 4 of 93 foragers collecting pollen. That too few low-strain bees become pollen foragers has been a constant difficulty for our foraging-comparison studies. High-strain pollen and nonpollen foragers were also significantly different in their responses to 30 percent sucrose solution. Differences between high-strain pollen and nonpollen foragers were probably due to effects of foraging. Differences between high- and low-strain nonpollen foragers most likely were due to genetic differences, but we could not exclude differences in foraging behavior because we did not know their foraging experiences before they were captured.

Foragers return to the nest, pass off their nectar loads, feed, and then go out again. We know that feeding bees to repletion after they return from foraging trips reduces their responsiveness to water and sensitivity to sugar. We hypothesized that the time foragers spend in the nest after returning, before they initiate their next trip, resets their responses to water and sugar. We collected high- and low-strain bees as they departed the hive and then subjected them to the proboscis extension response (PER) sucrose assay. Departing high-strain bees were more responsive to water and more sensitive to sucrose than departing low-strain bees. These results showed that at least the differences in PER responses of high- and low-strain foragers persisted throughout their foraging events. But differences still could be largely or entirely due to their foraging experiences rather than fundamental differences in sensory responses to water and sugar stimuli—the classic chicken-and-egg question (compare H1 with H3 in Figure 5.8).

We tested young workers for PER responses. We were frustrated that we could not separate the effects of foraging behavior from the potential genetic effects on responses. Perhaps genetic differences just put different bees in different foraging roles, and then the foraging experiences themselves modulated their stimulus-response systems (H1 or H2 in Figure 5.8). Or perhaps the genotype of a bee (her genetic complement) affects the response systems, which in turn affect their foraging decisions (H3 or H4). Tanya Pankiw took newly emerged high- and low-strain bees, marked them, and placed them in common hives. She monitored the hives to make sure that none of the bees she tested had begun foraging. She tested bees of three age cohorts: (1) within the first 48 hours of life, (2) within the first week of life, and (3) within the second week of life. High- and low-strain workers of these cohorts had lived together in the same hive environment and had not foraged; therefore, their responses could not be consequences of their foraging experiences. High- and low-strain bees of all three cohorts differed significantly in sucrose sensitivity, with high-strain workers always more sensitive; H1 and H2 were thereby eliminated as plausible explanations. High-strain bees 0 to 2 days old were 100 times more sensitive to sucrose than low-strain bees of the same age. (We subsequently showed that 4-hour-old high- and low-strain workers differ; see Section 5.6). Sensitivity of high-strain bees did not increase with age. However, sensitivity of low-strain bees did increase, which suggests that high-strain workers are developmentally more advanced. This is consistent with their early onset of foraging behavior. High-strain queens and drones are also more sensitive to sucrose than are queens and drones of the low strain, and they never forage; this suggests fundamental, genetically variable, neurophysiological differences in sensory-response-system tuning between high- and low-strain bees.

5.4.1.2 High- and Low-Strain Bees and Foraging Loads   Comparisons of the crop (honey stomach) contents of high- and low-strain foragers confirmed our expectations based on the hypothesized relationships between water responses and sucrose sensitivity and foraging behavior. A greater proportion of high-strain foragers returned with water and with lower concentrations of nectar, while a larger proportion of low-strain bees returned empty. High-strain bees with high responsiveness to water and high sensitivity to sucrose accepted a broader range of liquids. Low-strain bees with less sensitivity to sugar accepted only the higher-quality nectar or returned empty. Collectively, the results of this study caused us to reject the hypothesis that observed differences in sucrose sensitivity of high- and low-strain workers are simply consequences of foraging experience. Instead, the results support the hypothesis that genotype determines sensitivity to sucrose (and perhaps foraging behavior independently), that foraging behavior modifies it, and that the response is set very early in life and correlates with behavior much later (H2 or H6; Figure 5.8).

5.4.1.3 Sucrose Response Predicts Foraging Behavior in Wild-Type bees   A crucial test of our hypothesis that there is a causal relationship between sucrose response and foraging behavior, derived from the observed differences between high- and low-strain bees, would be to test preforaging-aged wild-type bees for their PER responses and predict their future foraging behavior. We took brood combs from two colonies containing wild-type queens (bees that came from commercial stocks but not from the high- or low-pollen-hoarding strains) that were naturally mated. Using genetic markers and sampling workers from these queens, we confirmed that they were mated to at least nine and five males, respectively, so each colony had a broad base of genetic diversity. Newly emerged workers were marked and placed in a single hive containing a naturally mated queen and bees that were unrelated to the test bees. Seven hundred marked workers were recaptured during their first week of life, taken into the laboratory, and tested for their response thresholds to sucrose. The response threshold is the lowest concentration for which the bee can distinguish sucrose solution from water. Each bee's threshold was recorded, a numbered plastic tag was glued to the thorax for individual identification, and the bee was returned to the hive. Tagged bees were collected after they initiated foraging, when they were more than 2 weeks old. We then compared the foraging loads they collected with their response thresholds measured weeks earlier. The results confirmed our hypothesis. Bees that collected water had the lowest response thresholds (highest sensitivity to water and sucrose), followed by bees that collected only pollen, then bees that collected only nectar, and bees that collected both pollen and nectar. The bees that returned empty were the least responsive group when they were tested at less than 1 week old (Figure 5.9). We found a strong correlation between nectar concentration and response thresholds. Bees that were less responsive (higher threshold) collected nectar of a higher sugar concentration. Response threshold explained a remarkable 53 percent of the total variance in nectar concentration. Tanya Pankiw and coworkers repeated the experiment two more times in different environments using different sources of bees, including Africanized honey bees, and obtained similar results.

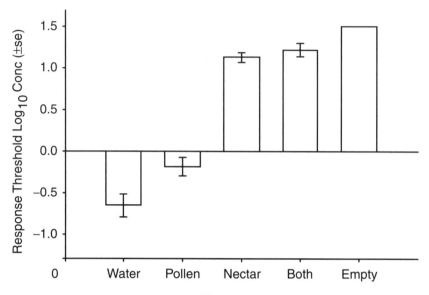

Figure 5.9. Average responsiveness of foraging honey bees returning with water, pollen, both water and pollen, or empty. Bees were tested when they were less than one week old and were then collected when they returned from foraging trips. Reprinted with kind permission from Springer Science+Business Media: *Behavioral Ecology and Sociobiology,* Response thresholds to sucrose predict foraging division of labor in honeybees, 47(4), 2000, 265–267, Pankiw T, Page RE, Fig. 1.

This is a robust finding, but it does not leave us with a single hypothesis. H4 and H6 can both explain the results that genotype affects sucrose response and sucrose response correlates with foraging behavior. (H3 and H5 fail because we know that foraging experience does affect response to sugar [RS].) Further resolution will require modulating RS directly and independently and then observing changes in foraging behavior, something we do not yet know how to do.

5.4.2 Sensitivity to Pollen

Sensitivity to pollen correlates with sensitivity to sugar. Pollen touched to the antenna can elicit a proboscis extension response in bees that are also sensitive to sucrose. Christoph Grüter and Walter Farina showed that wild-type pollen foragers were more sensitive to sucrose and also were more likely to exhibit a PER when pollen was touched to their antennae.

5.4.3 Sensitivity to Brood

High-strain bees are more sensitive to brood stimuli. Tanya Pankiw showed in common-garden experiments that high-strain bees responded more than low-strain bees when the pollen-foraging stimuli of stored pollen and larvae were changed. They increased their pollen-loading bias and foraged earlier in life when there were more larvae and less stored pollen. However, those studies confounded stimuli because both stimuli were changed simultaneously.

Jennifer Tsuruda held the amount of stored pollen constant in hives and looked at the effects of brood by having some colonies with a fixed amount of larvae and others completely broodless. Colonies contained cohorts of high- and low-strain workers, again a common-garden experiment. High-strain bees were more sensitive to the brood stimulus; the pollen-foraging bias increased when brood was present in the hive. High- and low-strain bees still differed in their pollen-foraging biases in the hives without brood, showing that pollen foraging is not a response only to brood stimuli, and that differences between highs and lows are not based only on responses to brood.

5.4.4 Sensitivity to Light

High-strain bees are more responsive to low levels of light. Within the high and low strains, light sensitivity correlates positively with sensitivity to sucrose. The same relationships are found with wild-type bees. Pollen foragers are more responsive to light and sucrose than are nectar foragers.

5.5 Associative Learning

High-strain bees perform better on tactile associative learning tests than do bees from the low strain. Pollen-foraging wild-type bees perform better than nectar foragers. The trials of conditioned stimulus

(CS)–unconditioned stimulus (US)–reward constitute the acquisition phase of learning. Bees vary in how many trials it takes to respond without the US (Figures 2.10 and 5.10). High-strain bees require fewer trials than low-strain bees. Pollen-foraging wild-type bees, like high-strain bees, require fewer trials than do nectar foragers. Following the acquisition trials comes the extinction phase, where bees are presented with the CS without reward. The bees learn to ignore the unrewarded stimulus. Bees are also tested for their ability to discriminate between two stimuli. In the case of tactile learning, they discriminate between a copper plate with vertical etchings and one with horizontal etchings.

Similar learning assays have been conducted using odors. Preforaging high-strain bees and pollen-foraging wild-type bees have better

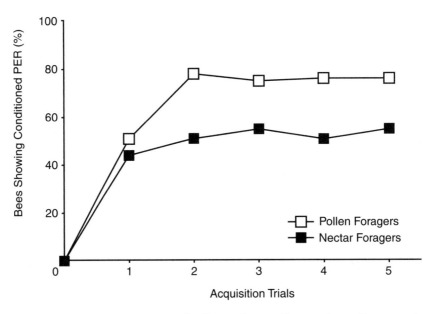

Figure 5.10. Acquisition curves of pollen and nectar foragers in tactile antennal learning. The x-axis shows the acquisition trials. The y-axis shows the percentage of bees displaying the conditioned proboscis extension response (PER). Both groups have reached the plateau of their acquisition function after three acquisition trials. However, the level of acquisition is higher in pollen foragers than in nectar foragers.

acquisition and have less extinction than preforaging low-strain and nectar-foraging wild-type bees. In the case of odors, an airstream with a training odor (the CS) is blown on the antennae for a brief time before the sugar solution is touched to the antenna (the US), followed by the sugar reward presented to tip of the proboscis. After very few trials, the bees learn to extend their proboscis when presented with the odor. Citral and carnation are typical odors used in olfactory associative learning.

Learning differences between pollen and nectar foragers and between preforaging-aged high- and low-strain bees can be attributed to their differences in sensitivity to sugar. This holds for both tactile and olfactory learning. The sucrose-response threshold of a bee provides a measure of her subjective value of the reward, as opposed to the true value determined by the caloric content. Bees that are more sensitive to lower concentrations subjectively value a reward of a sucrose concentration more and, as a consequence, perform better in associative learning. When bees are presented with equal subjective rewards, they perform the same whether they are high- or low-strain bees or pollen or nectar foragers. Equal subjective rewards are determined by presenting a concentration for a reward that is determined by the sucrose-response threshold of the bee. For example, if two bees vary in their response thresholds by one log unit (10-fold)—say, one responds at 1 percent, the other at 10 percent—then the less responsive bee will be tested with a concentration 10 times greater than that for the more responsive bee. In this case the less responsive bee will be tested with 30 percent, the more responsive bee with 3 percent.

The reward is evaluated at the proboscis. Ricarda Scheiner has shown very nicely that even though the proboscis is less sensitive to sucrose than are the antennae, the responses of the proboscis and the antennae are strongly correlated. This suggests that differences in sensitivity measured at either point are consequences of differences in central nervous system processing.

5.6 Nonassociative Learning

High- and low-strain bees also differ for elements of nonassociative learning. Nonassociative learning can take the form of sensitization or habituation. Sensitization occurs when a bee exhibits increased responsiveness following exposure to a strong stimulus. This is demonstrated when a bee that has been stimulated with a high-sucrose solution responds to a low-sucrose concentration that she would not respond to before the strong stimulus. A decline in response with repeated, unrewarded stimulation is habituation. For example, a bee that is repeatedly stimulated with water or a low-sucrose concentration gradually ceases to respond. Not only does sensitivity to sugar affect associative learning, but it also has an effect on nonassociative learning.

Vasantha Kolavennu worked in my lab in 1999 as a high-school intern and performed some very nice experiments demonstrating that newly emerged high- and low-strain bees differ in their sucrose-response thresholds and that response thresholds correlate with sensitization and habituation (unpublished results). In the first experiment, Kolavennu tested whether high- and low-strain bees differ in their responses to water and sucrose solutions when they are newly emerged, before they have social or foraging experiences. Responsiveness was measured as total gustatory responses (gustatory response score, GRS) for the seven solutions tested (water plus six sucrose solutions). Thus, a higher GRS signifies a lower response threshold (RT), while a lower GRS signifies a higher RT (Figure 2.7). There was a significant difference in the GRSs of newly emerged low- and high-strain bees; high-strain bees were significantly more responsive. This is interesting because the bees were just 4 hours old but already demonstrated differences in their sensory responses that correlated with foraging behavior weeks later.

Next, Kolavennu tested bees for their level of sensitization and resistance to habituation. After final stimulation with 30 percent sucrose solution, bees were presented with 10 trials of pure water to see how long they would continue to respond after being sensitized. Previous studies with Jochen Erber had shown that bees could be sensitized to respond to water after treatment with 30 percent sucrose solution. High-strain bees were significantly more responsive to water following sensitization and maintained a higher responsiveness to water throughout the process of habituation (Figure 5.11).

Finally, Kolavennu tested the effects of single and repeated stimulation with a 50 percent sucrose solution. Newly emerged high- and low-strain workers were exposed to 1, 3, and 6 stimulations with a 50 percent sucrose solution, respectively. Four minutes after stimulation, they were tested with 10 stimulus treatments of water touched to the antenna. The total number of responses demonstrated the degree of sensitization and resistance to habituation. The number of sucrose stimulations

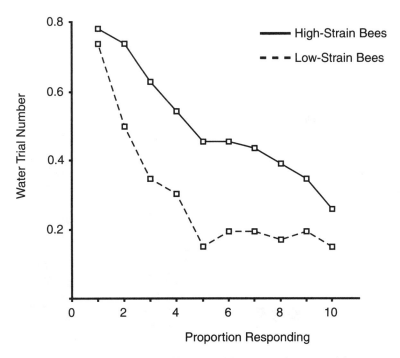

Figure 5.11. Habituation curves of high- and low-strain bees tested for response sensitivity using an increasing series of concentrations. The last sucrose concentration used to test for sensitivity was 30 percent. Afterward, bees were presented with 10 trials of water. The figure shows the proportion of highs and lows responding to 0 to 10 water trials of 92 bees.

strongly correlated with the total water responses, demonstrating that repeated stimulation with 50 percent sucrose increasingly sensitized the bees. High-strain bees in general were more sensitive to the first and subsequent stimulus trials (Figure 5.12).

These experiments clearly demonstrate that high- and low-strain bees differ in their responses to sucrose when they are newly emerged, before social or foraging experiences. Responsiveness to sucrose correlates with nonassociative learning, as it does with associative learning. This shows that they share a common mechanism that involves the subjective evaluation of stimuli. High- and low-strain bees are differentially sensitized by single and multiple exposures to sensitizing stimuli. The rates of habituation, desensitization, or both (we probably have elements of both in these assays) are related to sucrose sensitivity and differ between highs and lows.

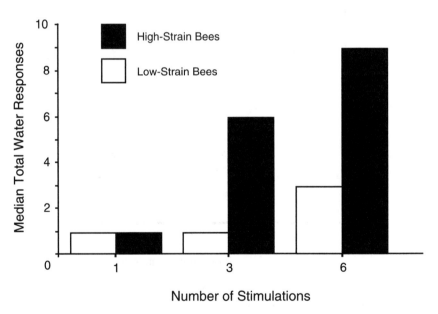

Figure 5.12. Median number of times 302 newly emerged worker honey bees of the high and low strains responded to 10 presentations of water stimulus to the antenna following 1, 3, and 6 successive presentations of 50 percent sucrose.

5.7 Motor Activity

Newly emerged high-strain bees are more active than low-strain bees. This can be seen in their walking activity soon after they emerge as adults. Merideth Humphries, a former graduate student of mine, took newly emerged adult high- and low-strain bees and placed them in a small arena made from a plastic petri dish. The arena floor was a piece of filter paper with a grid of squares drawn on it. Bees were placed in the arena and were observed for 2 minutes while they walked. The number of grid squares crossed by each bee was recorded as an indicator of its locomotor activity.

Humphries also looked at wild-type bees. After testing their locomotor activity, they were assayed for their responsiveness to sucrose, using the PER technique (Figure 2.7). Wild-type bees that were more active also were more responsive to sucrose. This significant correlation showed that the differences we observed between high- and low-strain bees (the coupling of locomotor activity and sucrose sensitivity) is characteristic of bees in general.

5.8 Neurobiochemistry

Behavior can be studied at many levels. One can sit in a blind and observe the behavior of animals, or one can look at the chemical changes occurring in individual cells of the brain associated with behavior as they interact. In one sense, the observational studies are like looking at an ocean beach at 10,000 meters elevation and seeing the shape of the beach, the whitecaps of the waves, and the contour of the cliffs, but not the grains of sand. On the other hand, studying behavioral neurobiochemistry is like studying the grains of sand one by one with a microscope and trying to understand what the beach looks like. When one considers the immense complexity of even the honey bee brain with its 900,000 neurons, one realizes how far we are from understanding behavior at this level. As a consequence, much of what we think we know today is certainly incomplete and probably wrong. But the hope is that as we generate more details about the operation of neurons and the biochemistry of their interactions, we will begin to put together a lasting map of the individual chemical and structural components of behavior and will begin to see the beach.

The neural system of insects is composed of the sensory receptors that detect tactile (touch), olfactory (smell), gustatory (taste), and light (vision) stimuli; interneurons that integrate electric signals from other neurons; motor neurons (also called motoneurons) that affect muscle contractions; and synapses that connect neurons to each other. Neurons are assembled into collectives called neuropils that form the anatomical structures of the brain (Figure 5.13). Using neurotransmitters, neurons within neuropils communicate with one another and with neurons located in different neuropils. Some neurotransmitters, such as the biogenic amines, act as neuromodulators because they change the transfer characteristics at synapses between neurons and because

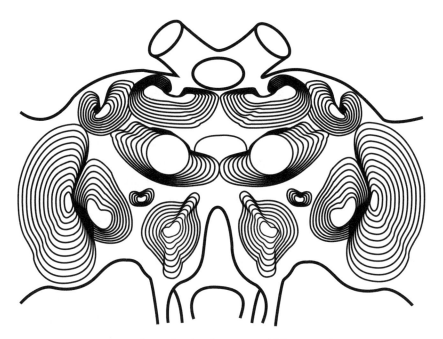

Figure 5.13. Drawing of a worker bee brain viewed from the front showing different structures that represent major neuropils (collections of neurons) that compartmentalize the brain.

they can change the intrinsic properties of cells, altering their sensitivities to inputs from other neurons. Insect biogenic amines are octopamine, tyramine, serotonin, and dopamine. Insect and human brains share serotonin and dopamine as important neuromodulators. Biogenic amines (also called bioamines) act by binding to proteins called receptors that are located on the surfaces of cells. There are many different receptors associated with the same bioamines that result in different consequences to the cells. Some bioamines cross-talk with (bind to) the receptors of others. When the bioamine binds to the receptor, it initiates an intracellular signaling cascade that involves so-called second messenger compounds, such as cyclic adenosine monophosphate (cAMP), cyclic guanosine monophosphate (cGMP), and protein kinases, such as cAMP-dependent protein kinase (PKA), calcium phosopholipid–dependent protein kinase (PKC), and cGMP-dependent protein kinase (PKG), that are activated and affect changes in other downstream signal molecules. It is obvious that the potential complexity of neural networks involved in behavior is enormous, with 900,000 neurons connected in complex ways involving multiple synapses and modulated by multiple bioamines, receptors, and cross-talking signal-transduction pathways.

We chose to look at a very tiny subset of the possible complexity of neurobiochemical pathways, namely, those that have been shown to be involved in sucrose sensory perception, motor activity, and learning and memory. Octopamine and tyramine, when fed to bees or injected into the thorax, increase responsiveness to sucrose. Dopamine decreases responsiveness when injected. Serotonin has no effect on sucrose response but does affect motor activity of the antennae. Dave Schulz, at the time a graduate student in the laboratory of Gene Robinson, looked at titers of the biogenic amines octopamine, dopamine, and serotonin in brains of high- and low-strain bees. He separated the brain into major regions reflecting different aggregations of neuropils. He looked at bees that were 1 day and 12 days old. Our expectation was that because high-strain bees are more responsive to sucrose as new adults, newly emerged and throughout life, they would have higher brain titers of octopamine and lower titers of dopamine. Because high-strain bees have higher motor activity, we expected them to have lower titers of serotonin. We found that 12-day-old bees had higher titers than 1-day-old bees, another prediction based on previous studies by Gene Robinson and his collaborators and consistent with increases with age in sucrose responsiveness, but the highs and lows did not differ from each other for any bioamine in any part of the brain, for either age group. Therefore, bioamine titers apparently are not directly responsible for differences in behavior of young high- and low-strain bees.

Behavioral differences between highs and lows could be attributed to other components of the signal pathways. Merideth Humphries found that high-strain newly emerged bees have higher brain levels of messenger RNA (mRNA) for the tyramine receptor *Amtyr1* (see Section 6.7 for an explanation of mRNA). If we assume that the amount of expressed *Amtyr1* mRNA is proportional to the amount of receptor protein that is active on the membranes of neurons, the tyramine receptor signal pathway could be more active in high-strain workers, which could perhaps explain some differences in behavior.

We looked at other genes in the signal cascade that are downstream from *Amtyr1*. High-strain bees contain more protein kinase A. PKA is the target of cAMP that is modulated by tyramine and octopamine. High-strain bees also have more protein kinase C (PKC) and protein kinase G (PKG), activation targets of cAMP, cGMP, phospholipase C, and nitrous oxide (NO). Sucrose stimulation results in activation of both PKA and PKC in the antennal lobes of the honey bee brain, and cGMP is likewise involved in sucrose perception. Therefore, it is plausible that at least some differences observed in behavior between high- and low-strain bees (and potentially nectar- and pollen-biased foragers) involve differences in these key neurobiochemical signal-pathway components.

5.9 Anatomy of Worker Ovaries and Vitellogenin

High-strain worker bees have more ovarioles per ovary than do low-strain workers. Ovarioles are the filaments that make up the ovary.

Each filament is a separate assembly line for making eggs (Chapter 7). Wild-type pollen foragers have more ovarioles than nectar foragers. Wild-type bees with more ovarioles are more sensitive to sucrose, forage earlier in life, have higher titers of vitellogenin, bias foraging toward pollen, and collect nectar with lower concentrations of sugar. High-strain workers have more vitellogenin circulating in the blood than do low-strain workers. The relationships of the ovary and vitellogenin to other phenotypic traits are discussed in more detail in Chapter 7.

5.10 Phenotypic Architecture of Males

Males do not forage and do not engage in most of the activities of worker honey bees; therefore, we have collected far less information about the behavior of high- and low-strain drones. However, the phenotypic traits we have looked at confirm what we have observed in workers. High-strain drones are more responsive to sucrose, initiate flight earlier in life, and demonstrate higher locomotor activity.

5.11 Phenotypic Architecture of Africanized Honey Bees

Tanya Pankiw studied Africanized honey bees (AHB) and compared their behavior with that of bees of European origin. African and European populations of honey bees separated about 1 million years ago and underwent independent evolution in relative isolation from each other until African bees were introduced into Brazil in 1956. Pankiw studied a population of AHB in Texas and compared them with commercial European honey bees. AHB were more sensitive to sugar, foraged earlier in life, demonstrated a stronger bias for collecting pollen, and collected nectar with a lower concentration of sugar. In other words, they demonstrated the high-pollen-hoarding syndrome of high-strain bees (Table 5.2).

Table 5.2 Association of traits in the pollen-hoarding syndrome

Trait	Selected strains		Wild-type commercial EHBs		Population comparisons	
	High strain	Low strain	WT pollen	WT nectar	AHBs	EHBs
Forage age	−	+	−	+	−	+
Pollen load weight	+	−	+	−	+	−
Nectar load weight	−	+	−	+	−	+
Nectar concentration	−	+	n/a	n/a	−	+
Gustatory response	+	−	+	−	+	−
Visual response	+	−	+	−	n/t	n/t
Locomotor activity	+	−	+	−	n/t	n/t
Tactile learning	+	−	+	−	n/t	n/t
Olfactory learning	+	−	+	−	n/t	n/t

Note: Comparisons are shown for workers from the high- and low-pollen-hoarding strains, for pollen and nectar foragers from commercial colonies that were not selected for pollen hoarding, and for populations of bees of European (EHB) and African (AHB) descent. + is greater than the alternative in comparison; − is less than; n/a designates that the comparison was not applicable; n/t designates that the comparison was not tested.

5.12 A Pollen-Hoarding Syndrome

Over the past 20 or more years, we have built a complex network of phenotypic correlations that range from complex social behavior to neurobiochemistry and more (see Chapters 6 and 8). However, one of the basic mantras of biology, taught to every student of statistics, and especially graduate students from their first day, is "Correlation does not imply causation." Bill Shipley in his book *Correlation and Causation in Biology* asserts that the opposite is almost always true: correlation does imply causation, but we just do not know the underlying causal structures. If we observe a statistically significant, systematic relationship between two things (variables), then we have ruled out the likelihood that this association is due to chance, and something must have caused it. Observation of correlated variables is often the beginning of unraveling the casual structures, as is the case for the phenotypic architecture of pollen hoarding.

We have gone beyond building simple correlations. We have looked at the association of traits in different genetic sources under different environmental conditions and have conducted controlled experiments with wild-type bees to build a robust architecture of traits related to the amount of pollen stored in the comb (Table 5.2; Figure 5.14). We have called the fundamental individual behavioral components the *pollen-hoarding syndrome*. When one steps back and looks at the associations of these traits, a few trends are apparent. First, they are general, not specific to artificially selected differences between high- and low-strain bees or artifacts of the selection program. They are also intimately

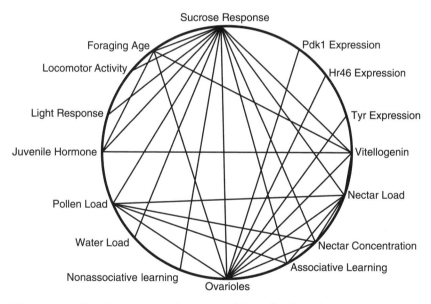

Figure 5.14. The phenotypic architecture of the pollen-hoarding syndrome. Phenotypic traits span levels of biological organization from the genotype to foraging behavior. Lines connect traits that have been demonstrated to be significantly correlated. Studies were performed on high- and low-strain workers, as well as wild-type bees. All traits shown here vary between bees of the high- and low-pollen-hoarding strains. Ovariole and vitellogenin phenotypic traits are discussed in detail in Chapter 7.

linked to the stimulus-response relationships best shown by the connections with sucrose sensitivity. This is not to say that differences in sucrose sensitivity cause the differences in foraging behavior—there could be common, unknown factors causing the variation in both and generating the correlations—but it does reveal the lack of independence of this constellation of phenotypic traits.

The dominant story that emerges from looking at Figure 5.14 is the linkage of traits. We selected for one thing, the amount of pollen stored in the comb, and changed a whole suite of phenotypic traits. These traits are linked causally because we find the same associations in wild-type bees. The possible phenotypic space, the character associations, is constrained, not free to vary independently for every trait. Selection is like squeezing a balloon to change its shape. You can squeeze and constrict it at one end, but it bulges out at the other, showing that the entire balloon is connected and interdependent.

What causes these associations? There are two plausible explanations. These traits could be linked together in a pleiotropic gene network where the effects of genes that vary between strains explain differences between high- and low-strain bees or between nectar and pollen specialists, or the suite of traits may be under broad hormonal control, and the high- and low-strain bees (and nectar and pollen specialists) may vary for hormonal signals that control the gene network, or both. I will explore those explanations in Chapters 6, 7, and 8. The network of associated traits reveals the architecture of the "spirit of the hive" through the networks that affect sensory-response systems and behavior.

Suggested Reading

Amdam, G. V., Ihle, K. E., and Page, R. E. 2009. Regulation of honeybee worker (*Apis mellifera*) life histories by vitellogenin. In *Hormones, Brain and Behavior*, 2nd ed., vol. 2, ed. D. W. Pfaff, A. P. Arnold, A. M. Etgen, S. E. Fahrbach, et al. San Diego: Academic Press, pp. 1003–1025.

Amdam, G. V., and Page, R. E. 2010. The developmental genetics and physiology of honeybee societies. *Anim. Behav.* 79:973–980.

Dreller, C. 1998. Division of labor between scouts and recruits: Genetic influence and mechanisms. *Behav. Ecol. Sociobiol.* 43:191–196.

Frisch, K. von. 1967. *The Dance Language and Orientation of Bees.* Cambridge, MA: Belknap Press of Harvard University Press.

Gordon, D. M., Barthell, J. F., Page, R. E., Fondrk, M. K., et al. 1995. Colony performance of selected honey bee (Hymenoptera: Apidae) strains used for alfalfa pollination. *J. Econ. Entomol.* 88:51–57.

Guzmán-Novoa, E., Page, R. E., and Gary, N. E. 1994. Behavioral and life-history components of division of labor in honey bees (*Apis mellifera* L.). *Behav. Ecol. Sociobiol.* 34:409–417.

Hellmich, R. L., Kulincevic, J. M., and Rothenbuhler, W. C. 1985. Selection for high and low pollen-hoarding honey bees. *J. Hered.* 76:155–158.

Humphries, M. A., Müller, U., Fondrk, M. K., and Page, R. E. 2003. PKA and PKC content in the honey bee central brain differs in genotypic strains with distinct foraging behavior. *J. Comp. Physiol. A* 189:555–562.

Hunt, G. J., Amdam, G. V., Schlipalius, D., Emore, C., et al. 2007. Behavioral genomics of honeybee foraging and nest defense. *Naturwissenschaften* 94:247–267.

Hunt, G. J., Page, R. E., Jr., Fondrk, M. K., and Dullum, C. J. 1995. Major quantitative trait loci affecting honey bee foraging behavior. *Genetics* 141:1537–1545.

Page, R. E., Amdam, G. V., and Rueppell, O. 2012. Genetics of reproduction and regulation of honey bee (*Apis mellifera* L.) social behavior. *Ann. Rev. Genet.* 46:97–119.

Page, R. E., and Erber, J. 2002. Levels of behavioral organization and the evolution of division of labor. *Naturwissenschaften* 89:91–106.

Page, R. E., Erber, J., and Fondrk, M. K. 1998. The effect of genotype on response thresholds to sucrose and foraging behavior of honey bees (*Apis mellifera* L.). *J. Comp. Physiol. A* 182:489–500.

Page, R. E., and Fondrk, M. K. 1995. The effects of colony-level selection on the social organization of honey bee (*Apis mellifera* L.) colonies: Colony-level components of pollen hoarding. *Behav. Ecol. Sociobiol.* 36:135–144.

Page, R. E., Scheiner, R., Erber, J., and Amdam, G. V. 2006. The development and evolution of division of labor and foraging specialization in a social insect. *Curr. Top. Dev. Biol.* 74:253–286.

Page, R. E., Waddington, K. D., Hunt, G. J., and Fondrk, M. K. 1995. Genetic determinants of honey bee foraging behaviour. *Anim. Behav.* 50:1617–1625.

Pankiw, T. 2003. Directional change in a suite of foraging behaviors in tropical and temperate evolved honey bees (*Apis mellifera* L.). *Behav. Ecol. Sociobiol.* 54:458–464.

Pankiw, T., and Page, R. E. 1999. The effect of genotype, age, sex, and caste on response thresholds to sucrose and foraging behavior of honey bees (*Apis mellifera* L.). *J. Comp. Physiol. A* 185:207–213.

Pankiw, T., and Page, R. E. 2000. Response thresholds to sucrose predict foraging division of labor in honeybees. *Behav. Ecol. Sociobiol.* 47:265–267.

Pankiw, T., and Page, R. E. 2001. Genotype and colony environment affect honey bee (*Apis mellifera* L.) development and foraging behavior. *Behav. Ecol. Sociobiol.* 51:87–94.

Rueppell, O., Bachelier, C., Fondrk, M. K., and Page, R. E. 2007. Regulation of life history regulates lifespan of worker honey bees. *Exp. Gerontol.* 42:1020–1032.

Rueppell, O., Fondrk, M. K., and Page, R. E. 2006. Male maturation response to selection of the pollen-hoarding syndrome in honey bees (*Apis mellifera* L.). *Anim. Behav.* 71:227–234.

Scheiner, R. 2004. Responsiveness to sucrose and habituation of the proboscis extension response in honey bees. *J. Comp. Physiol. A* 190:727–733.

Scheiner, R., Erber, J., and Page, R. E. 1999. Tactile learning and the individual evaluation of the reward in honey bees. *J. Comp. Physiol. A* 185:1–10.

Scheiner, R., Kuritz-Kaiser, A., Menzel, R., and Erber, J. 2005. Sensory responsiveness and the effects of equal subjective rewards on tactile learning and memory of honeybees. *Learn. Memory* 12:626–635.

Scheiner, R., Page, R. E., and Erber, J. 2001a. The effects of genotype, foraging role and sucrose responsiveness on the tactile learning performance of honey bees (*Apis mellifera* L.). *Neurobiol. Learn. Mem.* 76:138–150.

Scheiner, R., Page, R. E., and Erber, J. 2001b. Responsiveness to sucrose affects tactile and olfactory learning in preforaging honey bees of two genetic strains. *Behav. Brain Res.* 120:67–73.

Scheiner, R., Plückhahn, S., Öney, B., Blenau, W., et al. 2002. Behavioural pharmacology of octopamine, tyramine and dopamine in honey bees. *Behav. Brain Res.* 136:545–553.

Seeley, T. D. 1995. *The Wisdom of the Hive: The Social Physiology of Honey Bee Colonies.* Cambridge, MA: Harvard University Press.

Seeley, T. D., and Towne, W. F. 1992. Tactics of dance choice in honey bees: Do foragers compare dances? *Behav. Ecol. Sociobiol.* 30:59–69.

Shipley, B. 2000. *Cause and Correlation in Biology: A User's Guide to Path Analysis, Structural Equations and Causal Inference.* Cambridge: Cambridge University Press.

Tsuruda, J. M., and Page, R. E. 2009a. The effects of foraging role and genotype on light and sucrose responsiveness in honey bees (*Apis mellifera* L.). *Behav. Brain Res.* 205:132–137.

Tsuruda, J. M., and Page, R. E. 2009b. The effects of young brood on the foraging behavior of two strains of honey bees (*Apis mellifera*). *Behav. Ecol. Sociobiol.* 64:161–167.

Waddington, K. D., Nelson, C. M., and Page, R. E. 1998. Effects of pollen quality and genotype on the dance of foraging honey bees. *Anim. Behav.* 56:35–39.

Whitfield, C. W., Behura, S. K., Berlocher, S. H., Clark, A. G., et al. 2006. Thrice out of Africa: Ancient and recent expansions of the honey bee, Ap*is mellifera. Science* 314:642–645.

Winston, M. L. 1987. *The Biology of the Honey Bee.* Cambridge, MA: Harvard University Press.

— 6 —

The Genetic Architecture
of Pollen Hoarding

Underlying the linkage of phenotypic traits is a network of genes and developmental processes. In this chapter, I will show how genes responsible for variation in anatomical, physiological, behavioral, and social phenotypes have broad effects across levels and interact with each other. In Chapters 7 and 8, I will overlay developmental process on the building of the social phenotype.

6.1 Background

Soon after I left The Ohio State University and took a faculty position at the University of California–Davis, I received a call from Greg Hunt, a prospective graduate student. Hunt had received a master's in plant pathology and was working as a technician in a plant pathology lab at the University of Wisconsin–Madison. He was also a hobby beekeeper and wanted to do a genetic map of the honey bee. This was 1990: DNA marker techniques had only recently been developed for constructing genetic maps and used for mapping genes responsible for phenotypic traits. These techniques were being applied to plants, but little had been done with animals.

I was aware of some of the new genetic maps that had been constructed using restriction fragment length polymorphic (RFLP) DNA markers. The RFLP technique uses enzymes that cut DNA at places on the chromosome that contain specific nucleotide sequences, the building

blocks of DNA. The cuts allow DNA repair or transcription (the production of a translated copy of the DNA sequence of a gene that can be
used by the cell to make a protein or peptide) to take place. There are
many different enzymes that recognize many different sequences (cut
sites) scattered throughout the genome. Every individual has a unique
set of cut sites due to random mutations and the sequences inherited
from his or her parents. If you take the DNA of a single individual and
digest it with a specific restriction enzyme, the DNA will be cut into
many pieces of varying length, depending on where the cut sites are
located and the mutational history of the DNA segments they flank.
The digested DNA can then be run on a gel slab that separates the pieces
according to size, and they can be stained so you can see them. The
pattern observed gives information that can be used for mapping the
locations of the individual fragments in the genome. Greg wanted to
make a restriction site map of the honey bee.

Very little mapping had been done with honey bees. Before the development of DNA markers, genetic mapping was done with visible
mutations. The fruit fly *Drosophila* had an extensive map of mutations on chromosomes because of the large number of known visible
mutations and the large, stainable polytene chromosomes found in its
salivary glands. Even though instrumental insemination had made
the honey bee the model for studying social insect genetics, there
were few visible genetic markers and, therefore, only a few "linkage
groups" known, mostly the work of Harry Laidlaw and his colleagues.
When I left The Ohio State University, I went with the intention of
mapping the sex locus (see Chapter 4) once I was established at the
University of California–Davis. Hunt's phone call could not have
been timed better.

Hunt was admitted to graduate school and joined my lab in 1989.
About the same time, an article by a group of plant geneticists, some of
whom were at UC Davis, appeared in the journal *Nature*. They constructed a genetic map of quantitative trait loci (QTLs) for phenotypic
traits that had been bred in tomatoes. QTLs are genes that explain differences in phenotypes that must be measured (see Section 4.4.1.2). The
tomato map was one of the first QTL maps done by genetically map-

ping the entire genome and was very exciting. We wanted to map honey bee genes. We had the model, the expertise was available at UC Davis, and we were ready to go.

Hunt was one of those rare students who comes into a lab completely prepared technically and brings something new, defining a new direction for one's research. We discussed mapping QTLs for defensive behavior, a hot topic at the time because of the recent finding of Africanized honey bees in the United States. But I thought that it would be simpler and less risky (scientifically) to map pollen-hoarding behavior because we had the high- and low-strain bees. Truthfully, I thought that even that was too high a risk for a graduate student, even one as capable as Greg. (He did eventually go on to map defensive behavior as well.) The general belief at the time was that behavior was too complex to be able to map behavioral QTLs, especially a socially complex trait like stored pollen. We needed a sure-thing backup that could be mapped at the same time. The best trait would be something that has 100 percent penetrance (it is always expressed) and is a single gene—like the sex-determining locus. So we set out to map complementary sex determination and the pollen-hoarding trait simultaneously.

Mapping with RFLP markers is very laborious and slow. But our fortunes turned when Hunt's fiancée, Christie Williams, came to join him in Davis. A talented plant geneticist, she took a job in a plant genetics lab that was mapping plant genomes and became an important source for us of information on the newest markers and mapping techniques. Our biggest breakthrough came with the development of random amplified polymorphic DNA (RAPD) markers for mapping. Plant genetics labs at UC Davis were involved in the development of the markers and used them for mapping plant genomes. However, no one had attempted to use them to map an animal genome—until Hunt. Hunt worked with Williams and others in the plant genetic labs on campus to perfect the technique for honey bees. These new markers would greatly speed up our effort to map behavior in the honey bee.

6.2 Mapping Pollen Hoarding

There are three steps in making a QTL map. First, you make a genetic cross. Second, you construct a linkage map, showing how all your genetic markers are linked. Third, you do a QTL map on top of it. To make the map, you must make crosses where you have parents that differ with respect to the markers they have at any given point in the genome, and you know the marker that each parent has at each marker locus.

6.2.1 The Cross

We instrumentally inseminated a low-strain queen with the semen of a high-strain drone. From this cross we raised a new queen that was a hybrid of the two strains. The queen inherited one set of chromosomes from her low-strain mother and one set of chromosomes from her high-strain father. From her we raised more than 100 males. Males are haploid—they have no father. Therefore, each male had a unique genome that he inherited from his mother, a result of recombination during meiosis (Section 3.3 and Figure 3.4). We knew the origins (high or low strain) of all markers because we saved the drone father and included him in the marker analysis. We used 96 drones to construct the genetic linkage map. We chose 96 because that is the number of wells on the thermocycler we used. A thermocycler is a device that makes strands of DNA in a tube that fits into the well. DNA is made by a process called the polymerase chain reaction. The DNA produced contains the markers we use for constructing the maps. Note that today, mapping is done by direct sequencing of DNA, but the principles are the same.

The linkage map was derived from a total of 365 RAPD markers (Figure 6.1) assembled into 26 linkage groups of at least three markers. The overall size of the map was 3,450 centimorgans (cM; see Section 3.3). This is a huge recombinational map when it is compared with other animal species. The enormous size of the map is the reason we had 26 linkage groups when there are only 16 chromosomes for the honey bee. The higher the rate of recombination, the more markers you need for your map to become marker saturated and coalesce into the

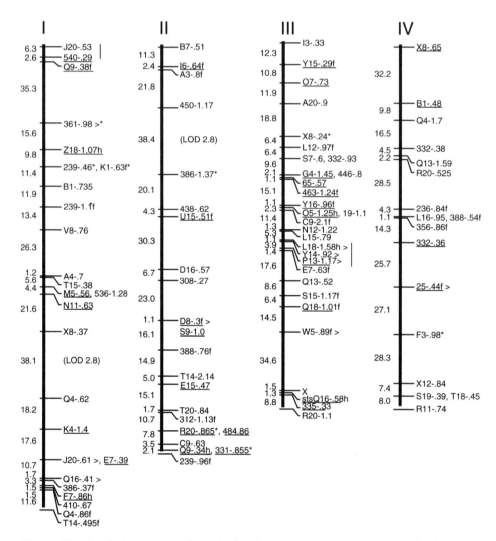

Figure 6.1. Four linkage groups from the first honey bee genomic map. The left side of each linkage group shows the distance in centimorgans (cM) between the markers shown on the right side. Markers were designated on the basis of the Operon® primer used to amplify them and their fragment size. Note that the complementary sex determiner *(csd)* discussed in Chapter 4 is mapped at the bottom of linkage group III, marked *X*. From Hunt G.J. and Page R.E. 1995. Linkage map of the honey bee, *Apis Mellifera,* based on RAPD markers. *Genetics* 139:1371–1382, Figure 1.

same number of linkage groups as chromosomes. This was already an enormous effort. To construct the map, we set up and ran about 30,000 individual polymerase chain reactions, a task that took more than two years.

6.2.2 Mapping QTLs

The next step was to map QTLs for pollen hoarding. Our map was sufficiently saturated even though our linkage groups did not coalesce into 16 chromosomes. We had a marker on average every 10cM, meaning that we had an average probability of recombination between our markers of 10 percent. But we needed phenotypes to assign to the genomes that constituted the map.

Males that were used for the map were also fathers of colonies (Figure 6.2). Males were randomly selected to provide sperm for instrumental insemination of high-strain queens. Before we ground them up to take their DNA, we ejaculated them and collected their sperm. The queens were super sisters, derived from the same mother who had been inseminated with the semen of a single drone. This is known as a high-strain backcross. We crossed the hybrid drones to high-strain queens because we had already shown that the pollen-hoarding trait, as well as individual foraging behavior and many other trait differences between highs and lows, shows directional dominance for the low-strain phenotype. In hybrid crosses, the expressed phenotypes are often more like the low strain, not intermediate in value (Figure 5.7). A backcross to a low-strain queen is not likely to show sufficient variation in phenotype to map the QTLs. It is like the inheritance of eye color. Genes (alleles) for blue eyes are recessive to the dominant brown eyes. An individual that inherits a gene for brown eyes from one parent and blue eyes from the other will still have brown eyes. If one parent has two alleles for blue eyes (one inherited from each parent) and the other parent has two brown alleles, all of their children will have one allele of each, and, therefore, all will have brown eyes—no phenotypic variation.

The high-strain queens that were inseminated with the sperm from the hybrid drones were placed in colonies where they laid eggs, and after

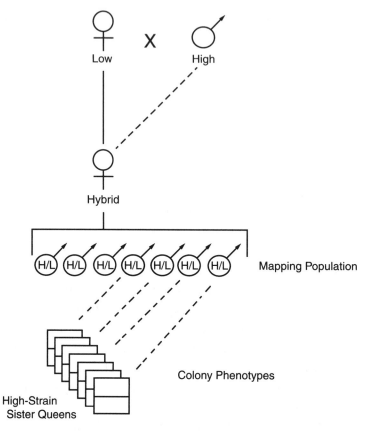

Figure 6.2. Males derived from a high/low hybrid queen were used to construct the linkage map in Figure 6.1. The same males were then used to instrumentally inseminate high-strain queens and hence became fathers of colonies that were used for the QTL map.

about 6 to 10 weeks, all the workers in their colonies were their offspring. We then evaluated 38 colonies for the amount of stored pollen and assigned the colony phenotype to the father of the colony. We assumed that because the queens were all super sisters derived from a breeding program with strong selection, they were fairly uniform for genes that affected the pollen-hoarding trait. They shared more than 87 percent of their genes in common. Therefore, the majority of the heritable variation we observed among colonies was due to the effect of the genetic contributions of the fathers.

Many computer programs now exist to construct linkage and QTL maps. However, the field was brand new when we were mapping pollen hoarding. We needed a program called MapMaker QTL that was developed by Eric Lander and his associates. Lander was leading the part of the Human Genome Project that was working to develop tools for constructing genetic maps and mapping QTLs. He visited UC Davis about the time we needed to begin building the map, installed his program on my computer, and showed us how to use it. Although the mathematical algorithms for mapping are complex, the general principle is easy. Think about doing an independent statistical test at each marker along one of the linkage groups shown in Figure 6.1. Start with marker J20.53 at the top of linkage group I and take all colonies that inherited the high-strain allele for that marker on one list and all those that inherited the low-strain marker allele on a different list. Then ask the question "Do the colonies represented on the two lists differ from each other for the amount of pollen stored in the comb?" If you do this successively for every marker, you are scanning the genome for regions that seem to be affecting the amount of stored pollen. However, you are doing hundreds of individual comparisons of markers that are not completely independent of one another because they are genetically linked. QTL mapping programs control for the large number of comparisons and genetic linkage.

Greg Hunt ran the program and found two major QTLs (Figure 6.3). We had succeeded where I had been certain we would fail. Perhaps the genetic basis of observed variation in naturally occurring social behavior wasn't that complex after all. We named the QTLs *pln1* and *pln2*. Together they explained 59 percent of the total variance in the pollen-hoarding phenotype. Contrary to what we expected, the low-strain allele for *pln1* resulted in more stored pollen than did the high-strain allele. This is a phenomenon that has persisted throughout our mapping, across multiple individual phenotypes, and is difficult to explain, but it occurs commonly in plant mapping studies where two selected populations are crossed.

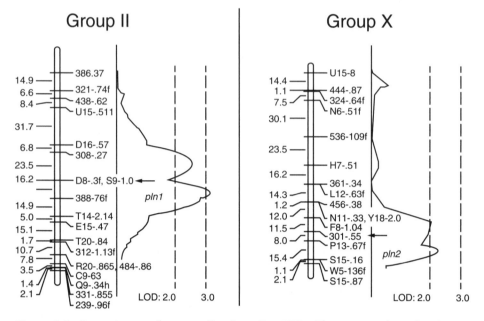

Figure 6.3. Genomic maps for two pollen-hoarding QTLs. The traces to the right of linkage groups map the regions of the genome according to the statistical probabilities that there are QTLs located in those regions. The LOD scores (Logarithm of Odds) are log-likelihood values. The higher the value, the more likely it is that the region contains a QTL. From Hunt et al. 1995, "Major quantitative trait loci affecting honey bee foraging behavior," *Genetics* 141:1537–1545, Fig. 2.

6.3 Verification of Quantitative Trait Loci

In QTL mapping, nothing should be considered real until it is independently verified. This is good practice in all the sciences, but especially in mapping QTLs for behavior, where we regularly read about a new gene "for" X, Y, or Z today, and then the next year someone publishes that he or she was unable to confirm it. We decided to confirm our pollen-hoarding QTLs independently by looking at the foraging behavior of individuals.

We mated a hybrid queen, derived from a fifth-generation cross, to a high-strain drone by instrumental insemination (Figure 6.4). The drone contributed identical genomes to all the workers in the colony because he was haploid and did not have recombination when he was

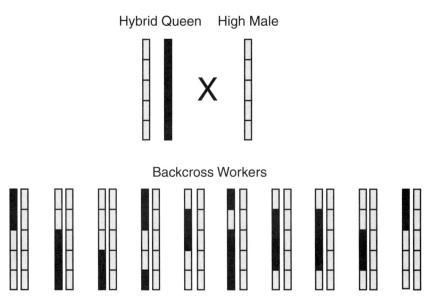

Figure 6.4. Mating design for validation of *pln1* and *pln2*. Worker genotypes were derived from a high-strain backcross. Just 1 set of chromosomes is shown, representing the full set of 16.

producing sperm gametes. However, every egg of the queen was a unique mixture of genes derived from her high and low parents. All workers inherited the same high-strain alleles for QTLs from their father but had a 50 percent chance of inheriting the high or low allele from their mother. As a consequence, heritable variation in the foraging behavior of individuals was derived from the mother.

We looked at the specific regions of the genomes of more than 320 returning foragers and determined whether they inherited a high or low QTL allele (as indicated by the RAPD marker allele) from their mother. Our expectation was that foragers that collected more pollen and less nectar would be more likely to inherit a high-strain QTL. We put the inseminated hybrid queen in a colony, marked 3,000 newly emerged workers (daughters of the queen), and collected them when they returned from foraging trips several weeks later. We weighed the pollen and nectar loads and determined the sugar concentration of nectar collected. Pollen and nectar loads correlated with our marker alleles, indicating that they correlated with our mapped QTLs. We had confirmed them. Consistent with the colony-level pollen-hoarding study, the high allele for *pln1* resulted in a bias for nectar, not pollen. We also looked at the effects of the QTLs on the sugar concentration of collected nectar. Bees that inherited the low-strain allele for *pln2* collected nectar with a significantly higher concentration of sugar. This result was initially very surprising, but subsequent studies that linked the genetic and phenotypic architectures of pollen hoarding with sucrose sensitivity and water and nectar foraging confirmed it (Chapter 5).

6.4 Identification of *Pln3*

We mapped two major quantitative trait loci that explained nearly 60 percent of the total phenotypic variance in our mapping population. The mapping population consisted of only 38 colonies derived from the second generation of our selection program. This is a very small number of individuals for a map. Normally you want many more. Merideth Humphries was appointed to do another map. Kim Fondrk set up the crosses from the fifth generation of the selection program using the same design as for the first map. This time the population consisted of 153 colonies headed by super-sister high-strain queens mated to hybrid drones. Phenotype data were taken 10 to 12 weeks after introduction of the queens to the hives. Two years later, the resulting QTL map revealed a new pollen-hoarding QTL that explained about 10 percent of the total phenotypic variance. We named it *pln3*. Like the *pln1* QTL, the high allele of *pln3* resulted in less stored pollen. Independent verification of *pln3* was done as for *pln1* and *pln2*.

During two sampling periods, we collected nearly 400 returning foragers derived from a high-strain backcross and examined them for marker alleles linked to *pln3*. The two sampling periods differed markedly with respect to the availability of floral resources, which was reflected in the results. Individuals that inherited the low-strain allele for *pln3* collected larger pollen loads in one sampling period, larger nectar loads in the other sample period, and nectar loads with lower sugar concentrations in both periods. Clearly *pln3* was affecting foraging behavior, and clearly its effects were dependent on the environment. It was also clear that the pollen-hoarding QTL must interact in a complex network with effects on multiple foraging traits.

6.5 *Pln4* and Mapping the Interactions of Pollen-Hoarding QTLs

Three pollen-hoarding QTLs had been identified and verified to affect individual foraging behavior. Two of the QTLs demonstrated effects that were counter to our expectations—the high hoarding and foraging effects were inherited from the low strain. We had two hypotheses to explain these results:

1. The alleles for *pln1* and *pln3* were fixed (identical within strains but different between them) by chance in the high and low strains, not by our selection. A high allele for each was fixed in the low strain, and a low allele for each was fixed in the high strain. Their overall phenotypic effects on the strains were negative, but we could not get rid of them. This can occur in a small breeding population, a consequence of genetic drift (see Section 4.2.2). But we did not think that this was the case because we had outcrossed (a breeding technique where one breeds selected stock to parents outside the selected breeding population in order to introduce new genetic material) and reselected and continued to get consistent results. Alleles with effects of 33 percent and 10 percent of the total phenotypic variance should be easy to select in or out of a breeding population. The outcross should have offered an opportunity to get the alternative alleles for the two QTLs into the strain gene pools and select out the alleles giving opposite effects.

2. The effects of the alleles were conditional on the other genes with which they interacted. In this case, the effect of the low allele acting on its own might be high pollen hoarding or foraging, but in combination with low (or high) alleles at other loci the effect could be toward the low-pollen-hoarding/foraging phenotype. This is a characteristic of epistatic interactions between genes known as allele-conditional effects.

We needed to look at the interactions of the QTLs in detail. This time Olav Rueppell, a postdoctoral fellow in my lab, took on the task.

Greg Hunt had successfully taken RAPD markers, cut the DNA products from gels, sequenced them, and made specific primers of high reliability. These primers amplified markers called sequence tag site markers. This greatly improved the quality of the genetic markers we were using to tag our QTLs. In addition, he produced a marker for the honey bee gene *Amfor,* which encodes an enzyme that was shown by Yehuda Ben Shahar, a student of Gene Robinson, to be involved in the onset of foraging. We thought that we might as well take a look at that region of the genome because we could easily make a marker. Hybrid virgin queens derived from a high- and low-strain cross were backcrossed to either a single high-strain or low-strain drone to produce high- and low-strain backcross workers. We collected them when they became foragers and uncovered a complex network of QTLs affecting the phenotypes we measured, including a QTL at or near *Amfor.* We designated the region around *Amfor pln4.* We had a fourth QTL. This does not mean that *Amfor* is responsible for the behavioral differences we observed; that is much more difficult to confirm. But it does mean that *Amfor* or something close to it is responsible for some significant amount of the behavioral variation observed.

We also constructed two new QTL maps from the low- and high-backcross workers. A new kind of DNA marker was used, amplified fragment length polymorphism (AFLP). This allowed a map to be constructed in less than six months. Several new QTLs were located in both maps for each of the traits. Some had individual effects, others only in interaction with other QTLs. Some QTLs were statistically significant; others were "suggestive," that is, statistically significant if one was looking only at that one marker, but not meeting the more rigorous requirements for significance when one was looking at many different (about 400) markers at the same time. Rueppell subsequently constructed QTL maps of the age of foraging onset (age of first foraging, AFF) and the responses of bees to sugar, with the same result. Collectively, several new QTLs were identified and confirmed. The take-home message from all these studies is that the QTLs we mapped represent complex interactive networks of genes with broad effects on the behavioral traits of our pollen-hoarding syndrome.

6.6 Mapping the Ovary and Juvenile Hormone
Regulation by Vitellogenin

I argue in Chapter 7 that the traits involved in the pollen-hoarding syndrome are linked through a network of genes and hormones that regulate reproduction. The next obvious question was whether the QTLs we
mapped for pollen hoarding and the suite of foraging traits associated
with pollen hoarding were also affecting ovary size. We produced a
hybrid queen and backcrossed her to a high-strain male, generating a
distribution of genotypes and ovary phenotypes. We collected returning foragers, weighed their nectar and pollen loads, dissected their
ovaries, and counted ovarioles. Bees that collected pollen had on average more ovarioles than those that collected only nectar or returned
empty.

We dissected newly emerged workers and counted their ovarioles.
Then we extracted their DNA to determine whether they inherited low-
or high-strain markers linked to each of the pollen-hoarding QTLs.
This was a fast and efficient way to look for candidate QTL effects on
ovarioles. We found direct effects of *pln2* and *pln3* and an interaction
effect of all four QTLs, thus demonstrating that the pollen-hoarding
QTLs were affecting ovariole number and interacting in a complex way,
just as they do for the foraging traits.

It has been a mark of our research program to verify our findings
with bees other than the select strains. Allie Graham, a student of Olav
Rueppell's, mapped QTLs for ovariole numbers in a population of bees
derived from crosses of Africanized (AHB) and European honey bees
(EHB). We prescreened colonies of EHB and AHB that we had at the
bee lab in Mesa, Arizona, and identified AHB colonies that had high
ovariole counts and European colonies with low counts. We produced a
hybrid queen and backcrossed her to a drone from the high-ovariole
Africanized queen. The resultant workers had the genes they inherited
from the hybrid queen mixed up in the egg gametes produced, but all
had a high-ovariole, AHB genome from the sperm of their father. This
is the same as the breeding scheme used for mapping the high- and
low-strain behavioral traits. Workers were collected, ovaries were dis-

sected, and marker inheritance was determined for markers located near the pollen-hoarding QTL previously found. Graham found that *pln1, pln2,* and *pln3* explained significant amounts of the variation in the number of ovarioles in the mapping-population backcross, thus confirming the effects of these QTLs on ovary size in a cross completely independent of the high- and low-strain bees.

In an additional QTL mapping study, Kate Ihle, a graduate student co-advised by Gro Amdam and me, investigated simultaneously the ovary size of workers (number of ovarioles) and the amount of juvenile hormone circulating in their hemolymph in response to vitellogenin knockdown (see Chapter 7). Juvenile hormone (JH) and vitellogenin (Vg) coregulate each other. Decreasing Vg results in an increase in JH, and vice versa. This has been shown to be true in general in honey bees; however, they do not coregulate each other in the low strain. Ihle mapped ovary size and JH response to Vg in a high-strain backcross worker population and found that *pln3* affected the regulation of JH by Vg (unpublished).

The architecture suggests a causal structure of selection on stored pollen, resulting in the substitution of alleles of genes that affect the size and development of ovaries that affect hormones that are part of the reproductive regulatory network that, in turn, affect sensory responses and behavior (Figure 6.5). But what are the genes represented by the QTLs we mapped?

6.7 Candidate QTLs

Three events in the very long history of honey bee research have greatly enabled the success of the honey bee as a genetics research organism. The first was the development of the removable-frame hive by Lorenzo Langstroth. This allowed beekeepers and researchers to remove, examine, replace, and exchange combs of honey, pollen, and brood. The second was the development of instrumental insemination, mostly attributed to my mentor and good friend for more than 25 years, Harry Laidlaw, and to Otto Mackensen, a geneticist for the U.S. Department of Agriculture. Instrumental insemination enabled researchers to control the matings of queens and do real genetic re-

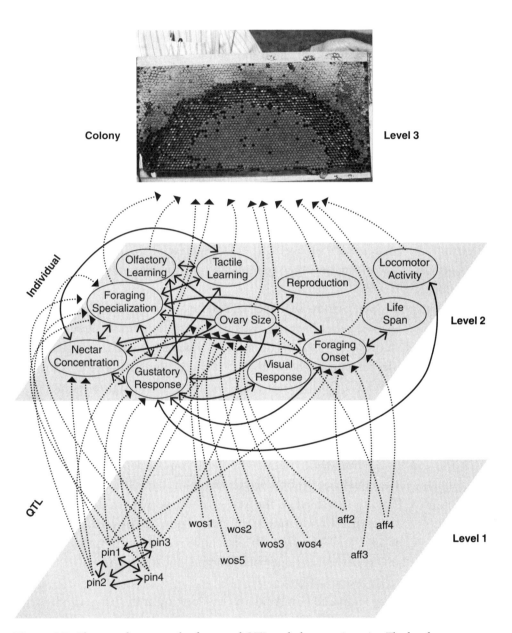

Figure 6.5. The complex network of mapped QTL and phenotypic traits. The levels shown are gene (QTL), individual (Level 2), and colony (Level 3). Dashed arrows show effects between levels. Solid arrows show within-level interactions (Level 1) and associations (Level 2). QTL traits are pln = pollen-hoarding QTL, wos = worker ovary size (number of ovarioles) QTL, and aff = age of first foraging (foraging onset) QTL, the same as foraging ontogeny. Phenotype traits: gustatory response is sucrose responsiveness; ovary size is ovariole number.

search. (It is interesting that the Austrian friar who discovered the fundamental principles of genetic inheritance, Gregor Mendel, worked with garden peas for seven years and began the field of genetics. He spent the rest of his life unsuccessfully trying to breed a better honey bee. He failed because he could not control the promiscuous mating of the queen.) The third was the completion of the honey bee genome sequence due to the huge effort of Gene Robinson. The genome sequence of the honey bee consists of more than 200 million base pairs (the building blocks of DNA), one-tenth the size of the human genome, and contains about 10,000 genes. The human genome contains about 20,000 genes. Besides opening up the honey bee to even more detailed and complete genetic analysis, the genome sequence has attracted geneticists working on other organisms into the community of bee research.

The honey bee genome sequence enabled us to search for positional candidate genes, genes in the regions of the chromosomes where our mapping studies of pollen-hoarding QTLs indicated that there is a gene or genes affecting our phenotypes. Statistical techniques allowed us to identify confidence limits around the locations on the chromosomes where our maps suggested that QTLs reside. We looked at the 97 percent confidence limits. This means that if you run the whole mapping experiment 100 independent times, 97 times the interval contained in the confidence limit will contain the gene you are looking for. We looked at the honey bee genome sequence within the confidence limits of our QTLs and identified all the known or suspected genes in those regions. Altogether we identified 116 candidates: 18 for *pln1*, 63 for *pln2*, 32 for *pln3*, and 3 for *pln4*. This is a huge reduction from the estimated 10,000 genes in the honey bee genome, but still too many to attempt to verify individually.

While Greg Hunt and Gro Amdam were looking at the list of candidate genes, they noticed that there was an overrepresentation of one set of genes that are involved in one signaling cascade. This cascade is known as the insulin-insulin-like signaling (IIS) pathway. It is ancient and ubiquitous, spread throughout the animal kingdom, including humans, insects, and worms. Signal cascades like the IIS network have many functions and are used in many ways. I think of them as door-bells. Many houses and buildings have them. When you ring the door-bell, the door opens, but you find very different things inside. Because of the broad effects of IIS pathways and their known effects on develop-ment, hormonal networks, reproduction, and metabolism, we decided to look at them more carefully.

The first step was to look at the expression of the different IIS genes in high- and low-strain bees. Genes are expressed when copies of them are being made in cells. However, the genes are made of deoxyribonu-cleic acid (DNA), but the expressed copies are made of ribonucleic acid (RNA). There are only subtle differences, but the differences make the RNA suitable as templates (messengers) for cells to make proteins (lon-ger sequences of amino acids) and peptides (shorter sequences of amino acids). If we found significantly different amounts of messenger RNA (mRNA) in high- and low-strain bees for specific genes, it could be an indication that those genes were affecting our phenotypes. If we didn't find differences, it would not mean that those genes are not the QTLs we mapped, just that the differences are not based on the current amounts of circulating mRNA. The genes themselves could be different with respect to structure and function of the proteins and peptides they encode. We looked at worker larvae, newly emerged adults, and forag-ers and found expression differences in two of six IIS genes. *PDK1* (*pln3*) was expressed at a higher level in the fat body of foragers from the high strain. *HR46* (*pln2*) was expressed at a higher level in low-strain bees of all life stages. *PDK1* and *HR46* were now candidate genes of special interest.

The next step was to see whether expression of *PDK1* and *HR46* var-ied with ovary size. We studied workers derived from a high backcross and looked for the association of gene expression and ovariole number that was presumably inherited from the hybrid queen, as we did in Section 6.3. We first looked at the association of ovariole number and foraging behavior. We found the normal associations, consistent with differences between high- and low-strain foragers: bees with more ovarioles were more likely to collect pollen, pollen foragers foraged ear-lier in life, and bees that foraged earlier in life for nectar collected nectar with lower sugar concentration. Newly emerged high-backcross workers with fewer ovarioles had higher expression of *HR46*, consistent with the pattern we saw in the expression study of highs and lows. There was no association of *PDK1* and ovary size in newly emerged bees, consistent with what we found in the expression studies. In foragers, *PDK1* was expressed more in bees with more ovarioles, as we found with high- and low-strain comparisons. These results demonstrated a genetic correlation in adult bees among *PDK1, HR46,* ovariole number, and foraging behavior. *HR46* probably acts early, perhaps on ovaries during larval development, and *PDK1* acts later, perhaps in response to differential signals from the ovaries.

6.8 Caveat

Genomics is still in an infant stage. The genome is hugely complex and interactive. Too often we try to reduce it to "a gene for X" when it clearly doesn't work that way. We have not yet figured out how to incorporate a view of the genome as part of the ecology of the gene, where the expression of genes and gene regulatory networks of the entire genome, combined with the internal physiology and external environment of the organism, are part of the effect of an individual gene. In Section 6.7, I presented one piece of what we think we know about the genomics of foraging behavior. I have not presented more because I am certain that what we think we know today will be wrong tomorrow, and then this book will already be out of date.

Suggested Reading

Amdam, G. V., Ihle, K. E., and Page, R. E. 2009. Regulation of honeybee worker (*Apis mellifera*) life histories by vitellogenin. In *Hormones, Brain and Behavior,* 2nd ed., vol. 2, ed. D. W. Pfaff, A. P. Arnold, A. M. Etgen, S. E. Fahrbach, et al. San Diego: Academic Press, pp. 1003–1025.

Amdam, G. V., Nilsen, K. A., Norberg, K., Fondrk, M. K., et al. 2007. Variation in endocrine signaling underlies variation in social life history. *Am. Nat.* 170:37–46.

Amdam, G. V., and Omholt, S. W. 2003. The hive bee to forager transition in honeybee colonies: The double repressor hypothesis. *J. Theor. Biol.* 223:451–464.

Amdam, G. V., and Page, R. E. 2010. The developmental genetics and physiology of honeybee societies. *Anim. Behav.* 79:973–980.

Ben-Shahar, Y., Robichon, A., Sokolowski, M. B., and Robinson, G. E. 2002. Influence of gene action across different time scales on behavior. *Science* 296:741–774.

Beye, M., Gattermeier, I., Hasselmann, M., Gempe, T., et al. 2006. Exceptionally high levels of recombination across the honey bee genome. *Genome Res.* 16:1339–1344.

Hunt, G. J., Amdam, G. V., Schlipalius, D., Emore, C., et al. 2007. Behavioral genomics of honeybee foraging and nest defense. *Naturwissenschaften* 94:247–267.

Hunt, G. J., Page, R. E., Jr., Fondrk, M. K., and Dullum, C. J. 1995. Major quantitative trait loci affecting honey bee foraging behavior. *Genetics* 141:1537–1545.

Iltis, H. 1932. *Life of Mendel.* New York: W. W. Norton.

Laidlaw, H. H. 1987. Instrumental insemination of honeybee queens: Its origin and development. *Bee Wld.* 68:17–36, 71–88.

Laidlaw, H. H., and Page, R. E. 1997. *Queen Rearing and Bee Breeding.* Cheshire, CT: Wicwas Press.

Page, R. E., Amdam, G. V., and Rueppell, O. 2012. Genetics of reproduction and regulation of honey bee (*Apis mellifera* L.) social behavior. *Ann. Rev. Genet.* 46:97–119.

Page, R. E., Erber, J., and Fondrk, M. K. 1998. The effect of genotype on response thresholds to sucrose and foraging behavior of honey bees (*Apis mellifera* L.). *J. Comp. Physiol. A* 182:489–500.

Page, R. E., and Fondrk, M. K. 1995. The effects of colony-level selection on the social organization of honey bee (*Apis mellifera* L.) colonies: Colony-level components of pollen hoarding. *Behav. Ecol. Sociobiol.* 36:135–144.

Page, R. E., Fondrk, M. K., Hunt, G. J., Guzmán-Novoa, E., et al. 2000. Genetic dissection of honeybee (*Apis mellifera* L.) foraging behavior. *J. Hered.* 91:474–479.

Page, R. E., and Laidlaw, H. H. 1988. Full sisters and super sisters: A terminological paradigm. *Anim. Behav.* 36:944–945.

Page, R. E., and Laidlaw, H. H. 1992. Honey bee genetics and breeding. In *The Hive and the Honey Bee.* Rev. ed. Hamilton, IL: Dadant and Sons, pp. 235–267.

Page, R. E., Waddington, K. D., Hunt, G. J., and Fondrk, M. K. 1995. Genetic determinants of honey bee foraging behaviour. *Anim. Behav.* 50:1617–1625.

Rueppell, O., Chandra, S. B. C., Pankiw, T., Fondrk, M. K., et al. 2006. The genetic architecture of sucrose responsiveness in the honey bee (*Apis mellifera* L.). *Genetics* 172:243–251.

Rueppell, O., Fondrk, M. K., and Page, R. E. 2006. Male maturation response to selection of the pollen-hoarding syndrome in honey bees (*Apis mellifera* L.). *Anim. Behav.* 71:227–234.

Rueppell, O., Metheny, J. D., Linksvayer, T., Fondrk, M. K., et al. 2011. Genetic
    architecture of ovary size and asymmetry in European honeybee workers.
    Heredity 106:894–903.
Rueppell, O., Pankiw, T., Nielson, D. I., Fondrk, M. K., et al. 2004. The genetic
    architecture of the behavioral ontogeny of foraging in honey bee workers.
    Genetics 167:1767–1779.
Rueppell, O., Pankiw, T., and Page, R. E. 2004. Pleiotropy, epistasis, and new
    QTL: The genetic architecture of honey bee foraging behavior. *J. Hered.*
    95:481–491.
Tanksley, D. 1993. Mapping polygenes. *Ann. Rev. Genet.* 27:205–233.

— 7 —

Reproductive Regulation of
Division of Labor

The view of the genetic, physiological, and behavioral mechanisms of social organization has become progressively more detailed and complex throughout Chapters 5 and 6. The model is full of ingredients, each apparently essential, but without an understanding of how they interact to affect the social phenotype, the flavor—not Mulligan stew, but certainly no longer stone soup. I needed to look for a more simplified explanation, one that can be more easily understood. In particular, the phenotypic architecture of pollen hoarding needed an explanation. Genetic mapping revealed an underlying genetic architecture where multiple genes, in this case mapped as QTLs (see Chapter 6), interacted with one another and had effects on multiple phenotypic traits. This suggested a broad, integrating regulatory network. But I was stumped. It was a twist of fate that led to the discovery that the network of genes and hormones that regulate reproduction had been hijacked and exploited by natural selection, and by me with high- and low-pollen-hoarding selection, and used to regulate foraging behavior.

7.1 Background

In 2003, I was asked to be an opponent for the doctoral dissertation defense of Gro Amdam at the Norwegian Life Sciences University. Universities in Scandinavian countries require that an external reviewer take part in the final decision on the suitability of the work submitted by doctoral candidates. In their system, however, the external reviewer is called an opponent and plays a special role in the process. I agreed and soon afterward received Amdam's dissertation. My role was to read it and develop questions to ask her in a public defense. The central part of her dissertation examined how juvenile hormone and vitellogenin coregulate each other and as a consequence regulate the onset of foraging behavior in bees. She and her doctoral advisor, Stig Omholt, had worked out an elegant model for how it can work and supplied some critical data in support of it. They called their model the double-repressor hypothesis.

7.2 The Double-Repressor Model

The double-repressor model proposes a new role for vitellogenin (Vg) in the life history of honey bees and a new interaction between vitellogenin and juvenile hormone (JH). In most insects, juvenile hormone signals the production of vitellogenin by globular structures in the hemolymph that are attached to the wall of the abdomen collectively called the fat body, and prepares the ovaries to incorporate Vg (Figure 7.1). (Insects do not have a closed, vascular system for moving blood around. They have an open system where the organ tissues float in the blood, called hemolymph.) The fat body serves the role of the liver and biosynthetic factory for insects. Vitellogenin is released from the fat body into the hemolymph, binds to the ovaries, is incorporated, and is used to make eggs. Amdam suggested that instead of just initiating the production of Vg, JH also suppresses it. But Vg also suppresses JH. After the initial surge in blood titer of JH, Vg production begins and suppresses JH. As Vg is fed to larvae, Vg titers decrease, and Vg loses its suppression of JH. As JH titer increases, it drives down Vg even faster, resulting in a sharp switch where Vg goes low and JH goes high (Figure 7.2). Workers then initiate foraging in response to the changes. Amdam and her colleagues were able to demonstrate that when Vg titers were suppressed by interference with production of Vg, JH titers increased, and bees initiated foraging earlier in life.

Sometimes clarity really does come in the shower. After reading the dissertation of Gro Amdam, I felt that the double-repressor model must

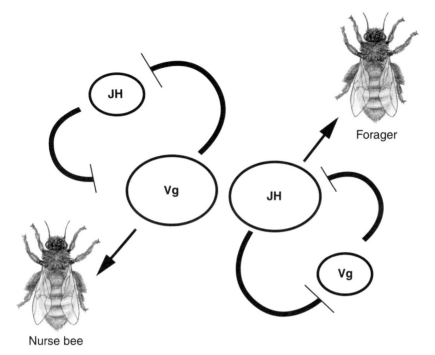

Forager

Nurse bee

Figure 7.1. In the double-repressor model, vitellogenin (Vg) and juvenile hormone (JH) are linked in a feedback system with a mutual ability to suppress each other. The nurse bee state (left) is governed by elevated vitellogenin titers that suppress JH and foraging behavior. Foraging is initiated by a worker when inhibition by vitellogenin is lost and JH levels increase, causing an additional reduction of vitellogenin levels via the regulatory feedback loop (right). Reprinted from *Hormones, Brain and Behavior,* 2nd edition, Vol. 2. Amdam GV, Ihle DE, Page RE, "Regulation of Honeybee Worker (*Apis mellifera*) Life Histories by Vitellogenin," in: Donald W. Pfaff, Arthur P. Arnold, Anne M. Etgen, Susan E. Fahrbach and Robert T. Rubin, editors. San Diego: Academic Press, 1003–1025, Fig. 4 (2009), with permission from Elsevier.

be the key to understanding the phenotypic and genetic architectures of pollen hoarding. The age of onset of foraging is a key component of the architecture (Figure 5.14), and she was able to link it to reproductive physiology through vitellogenin. It was in the shower that I remembered my Entomology 102 Insect Physiology course, where my professor, Charles Judson, taught us about the gonotropic cycle of mosquitoes. Mosquito females change their sensitivity to stimuli and their behavior

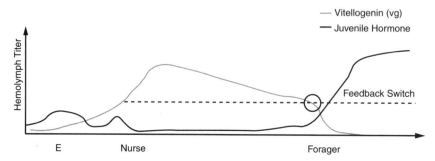

Figure 7.2. Blood titers of vitellogenin (Vg) and juvenile hormone (JH) in worker honey bees as they age from emerging as an adult (E) through nurse to forager. Reprinted from *Current Topics in Developmental Biology*, Page et al., "The Development and Evolution of Division of Labor and Foraging Specialization in a Social Insect (*Apis Mellifera* L.)," 253–286, Fig. 7 (2006), with permission from Elsevier.

according to the states of their ovaries, whether they are making eggs, have eggs, or have just laid eggs. I also remembered a paper by Mary Jane West-Eberhard where she wrote about the ovarian cycle in a primitively social wasp and how she thought that it was involved in reproductive division of labor. We needed to look at the relationships between the ovary and behavior, a relationship that is ubiquitous in insects and indeed throughout the animal world. I e-mailed Amdam and told her to be prepared to make the connections between her work and mine.

I was not disappointed. Amdam was totally prepared for my questions about her dissertation work and had worked out a tentative model of reproductive regulation of division of labor and an explanation of how her double-repressor model was linked to the pollen-hoarding architecture. She was awarded her degree, and a few months later she joined my lab to begin the search for empirical evidence. We quickly mapped out a set of experiments.

7.3 The Reproductive-Ground-Plan Hypothesis and Early Experiments

Amdam called the hypothesis the "reproductive-ground-plan hypothesis," a variation on the ovarian-ground-plan hypothesis put forth by

Mary Jane West-Eberhard. The original manifestation postulated that high- and low-strain bees were frozen in different parts of a gonotropic cycle of ancient origins in insects (Figure 7.3). Our thoughts were about the mosquito, where newly emerged adults forage for nectar from flowers to sustain themselves with carbohydrates for energy and then switch to foraging for protein by seeking blood meals. Associated with this is a change in behavior and presumably a change is sensory sensitivity. As protein foragers, mosquitos fly at dawn and dusk and orient to infrared radiation and carbon dioxide. After females take a blood meal, they seek low, dark places where they sit, digest their food, and make vitellogenin and eggs. When they have an egg load, they fly again and seek water vapor, where they lay their eggs on the water. Then they begin the cycle again. Our original, naive idea for the honey bee was that nectar foragers were stuck in the carbohydrate self-maintenance part of the cycle, while pollen foragers were stuck in the reproductive, protein-foraging part. It is important to note that many insects apparently do not cycle; instead, they display a sequence of ovary maturation with accompanying changes in behavior with age.

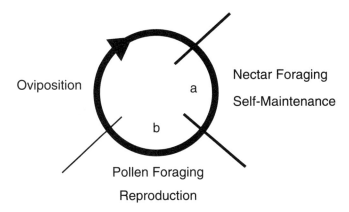

Figure 7.3. Early conceptual model of the reproductive-ground-plan hypothesis. Many insects change behavior as they cycle through reproductive states involving the availability of eggs in the ovary. Oviposition resets the cycle. We hypothesized for the honey bee that nectar foragers are stuck in stage *a*, while pollen foragers are stuck in *b*. Although the actual situation is certainly not this simple, the model provided testable hypotheses.

Our first prediction was that there should be differences in the dynamics of vitellogenin between highs and lows. We reasoned that because Vg was important for the onset of foraging, and there were clear, big differences in the age of foraging onset between highs and lows, there should be differences in Vg dynamics. We sampled newly emerged high- and low-strain bees and placed others in wild-type colonies to age for 5 and 10 days. In the sample bees, we looked at Vg protein titers in the blood and also at *vg* messenger RNA (mRNA) titers. The mRNA titers show how much transcription of the vitellogenin gene is taking place. We found that mRNA and Vg protein correlated, as expected, over the 10 days. Vg protein and *vg* mRNA levels were higher in high-strain bees than in lows, as predicted.

Next, we hypothesized that high-strain bees should be in a higher state of ovary activation than bees from the low strain and should develop their ovaries more rapidly in the absence of the queen because they were already in a higher reproductive state. In the absence of the inhibiting effect of the queen, worker ovaries develop, and some will lay eggs (see Section 3.9). Our first test compared ovaries from newly emerged high- and low-strain workers. We found that workers from the high strain were much more likely to have ovaries that had been activated and were previtellogenic: opaque, somewhat swollen ovarioles, unlike the clear, thin filaments of ovaries that are not activated (see Figure 7.4).

We employed an undergraduate, Angela Csondes, to work in my lab that summer. She was trained to do the dissections and score the ovaries. She commented that she thought that the high-strain bees had larger ovaries than the lows. She was right. Dissections showed that the high-strain bees she examined had on average six ovarioles, while the lows on average had only three. We also noted a correlation between ovariole number and ovary activation. Bees with more ovarioles were more likely to have them activated. This was true even for newly emerged bees (Figure 7.4). We then put newly emerged high- and low-strain workers into colonies with and without queens and dissected them 10 to 21 days later. In colonies with a queen, we found that about 30 percent of the high-strain workers had activated ovaries, compared

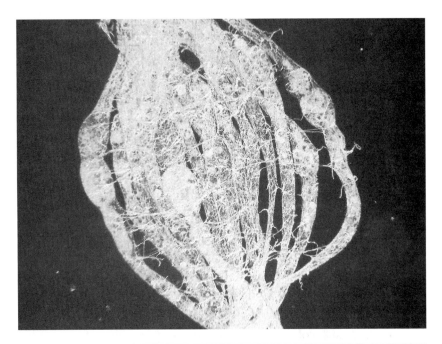

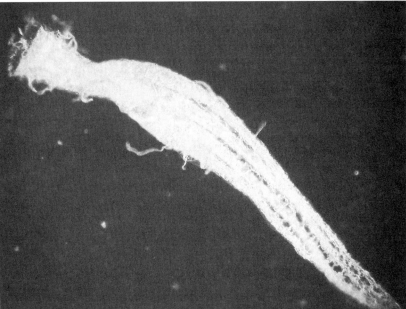

Figure 7.4. Ovaries from high- and low-strain workers. Bees had just emerged as adults and demonstrated significant differences in the stages of ovary development. The low-strain worker (lower panel) has fewer ovariole filaments, while the high-strain ovarioles (upper panel) are already swollen and opaque, a sign of reproductive activation. Photo by Osman Kaftanoglu.

with only 3 percent of low-strain bees. Only bees with seven or more ovarioles had them activated. In queenless colonies, 75 percent of high-strain and 42 percent of low-strain bees had activated ovaries. Forty-five bees had vitellogenic ovaries with eggs, the highest state, suggesting that these bees were active laying workers. Of the 45, 36 came from the high strain. This showed that the high-strain workers were at higher states of reproductive activation.

Because of Csondes's discovery of the different ovariole numbers in high- and low-strain workers, we now had a way to validate our hypothesis with wild-type bees. If our hypothesis is correct, then wild-type bees with more ovarioles should have ovaries in higher activated states (be more reproductively developed) and demonstrate a stronger bias toward collecting pollen. We took combs from four unrelated wild-type colonies that contained naturally mated queens. Because of the polyandrous mating of queens, each colony should contain a large number of subfamilies derived from the many mates of the queens and be genetically very diverse. Emerging adult workers were marked with paint, and the workers from the four sources were mixed together and put back in each of the four hives from which the mixture was derived. Marked bees were collected at the entrances over several days after they initiated foraging. Pollen loads were weighed, nectar loads were measured, and the concentration of the nectar collected was determined using a handheld refractometer. Our hypothesis of the relationship between ovary size and foraging behavior was confirmed. Bees with more ovarioles behaved like the high-strain bees: they foraged earlier in life and were more likely to collect pollen, and when they collected nectar, they accepted nectar with a lower concentration of sugar (see Section 5.12).

7.4 How Vitellogenin Affects Onset of Foraging and Foraging Behavior

We hypothesized that vitellogenin (Vg) was pacing the onset of foraging, perhaps through its suppression of juvenile hormone (JH) and probably other mechanisms. It has long been known that topical application of JH or analogs, chemical compounds with similar structure and function, results in early foraging. The JH could act directly on the physiology of the bee or indirectly through its effect on Vg. The same can be said about Vg: it could have JH-independent and dependent effects. Juvenile hormone is produced in the corpora allata of bees. These are paired secretory glands located on either side of and below the brain. Gene Robinson's student Joe Sullivan surgically removed the glands from bees and found that they still initiated foraging at close to the same age as bees that had not been surgically treated. He thus demonstrated that JH is not necessary for initiation of foraging. There must be a JH-independent pathway or pathways acting.

We also hypothesized that Vg cannot be acting directly on foraging decisions for pollen, nectar, or water because most Vg is stripped from the bodies of bees before they initiate foraging, it must instead be priming the bees for pollen foraging. (However, recent results show that high- and low-strain foragers do have blood titers of Vg, and high-strain bees have significantly higher levels than low-strain bees, perhaps directly affecting foraging behavior [Amdam, Ihle, and Page unpublished data].) We developed a model that incorporated both the onset of foraging driven by the double-repressor action of Vg and JH and foraging behavior using Vg as a primer (Figure 7.5). We propose that the onset of foraging corresponds to a reduction of blood titers of Vg that reduces the suppression of JH. JH in turn increases in titer and suppresses production of Vg. This leads to a rapid loss of Vg and the onset of foraging. However, high-strain bees have more Vg but forage earlier in life. Where does the Vg go so fast? The double-repressor model of Gro Amdam and Stig Omholt proposes that the blood-circulating titer of Vg is reduced by feeding larvae, queens, and other workers. Feeding larvae is the main source of loss. If that is the case, do high-strain bees perhaps engage in more brood rearing?

The titer of Vg in a worker rises over the first 10 or so days of life. We proposed that it must reach some threshold level before bees are primed to forage for pollen. However, it is more likely a continuum than a threshold because genetic variation exists in populations for Vg titers, as well as the way the bees load pollen and nectar when they forage. Some bees load more pollen and less nectar, others more nectar and

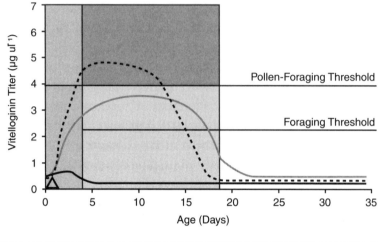

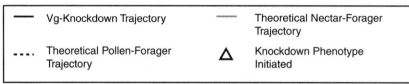

| — | Vg-Knockdown Trajectory | — | Theoretical Nectar-Forager Trajectory |
| .... | Theoretical Pollen-Forager Trajectory | △ | Knockdown Phenotype Initiated |

Figure 7.5. A model for the action of Vg in foraging division of labor. Very young bees are unable to forage. They must pass through an initial maturation phase during which flight muscles develop and the cuticle hardens. During this phase (gray) workers are primed for their future foraging specialization by titers of the protein vitellogenin (Vg). Vg titers above the pollen threshold prime workers for pollen foraging (dark gray), while workers with lower preforaging titers (light gray) are primed for nectar foraging. In workers, Vg suppresses the transition from nest tasks to foraging activity when its titer remains above the foraging threshold level. Below this threshold, the probability of initiating foraging is increased. This model explains the Vg-knockdown phenotype observed by Nelson and colleagues (Section 7.5.2). Ribonucleic acid interference-mediated knockdown of vitellogenin results in workers that mature with vitellogenin titers that are below both the pollen and foraging thresholds. These workers forage precociously and preferentially collect nectar. From Nelson et al. 2007. "The gene vitellogenin has multiple coordinating effects on social organization." *PLoS Biology* 5(3): e62, doi: 10.1371/journal.pbio.0050062, 673–677, Fig. 5.

less pollen (Figure 2.12). This relationship is heritable and varies considerably within and between populations.

7.5 Evidence for the Reproductive-Ground-Plan Hypothesis

Evidence for the reproductive-ground-plan hypothesis was building, but we had many skeptics. Criticism came in basically four flavors: (1) We really cannot understand the evolutionary history of social behavior by studying honey bees because they are socially too advanced, and the evolutionary history has probably been erased by 50 million years of social evolution. We dismissed this criticism as too pessimistic and continued to look for the signature of natural selection on the reproductive regulatory networks. That was the purpose of this research enterprise. (2) We know nothing about how reproductive state and behavior are coupled in closer relatives of bees. We considered this a valid criticism and an important question, but certainly not fatal to our hypothesis because the relationship between ovary state and behavior is widespread in animals. (3) Support for our hypothesis was derived from artificially selected strains and probably does not reflect what goes on in "normal" bees. This was just wrong because we had substantiated our results with highs and lows along with wild-type bees. (4) Critics just don't believe that there is a relationship between vitellogenin and foraging behavior . . . just because they don't believe it. This was difficult to dismiss because it was outside the context of the scientific method and was based on belief rather than evidence.

7.5.1 Ovaries, Vitellogenin, and Behavior

We made the argument that the amount of vitellogenin circulating in the blood of young worker honey bees affects their foraging behavior later in life. Workers are stripped of most of their Vg when they become foragers, so Vg titers are not likely to be directly affecting foraging decisions. We also argued that the amount of circulating Vg is affected by the size of the ovary. One mechanism for this could be the positive correlation between ovary size and ecdysteroid levels. Ecdysteroids are hormones that act on the fat body and stimulate Vg production. The causal chain in this case is clear: ovary → Vg → behavior. If this is true, then we should observe this correlative structure in wild-type bees.

Jennifer Tsuruda asked this question as part of her doctoral dissertation. She sampled 6–7-day-old wild-type workers from the brood combs of a colony and tested them for a relationship among ovary size, *vg* mRNA, and responsiveness to sugar, using the proboscis extension reflex (PER) test (Figure 2.7). Sucrose response correlates with the other behavioral traits in the pollen-hoarding syndrome (Chapter 5). She found that bees with more ovarioles had higher titers of *vg* mRNA and were more responsive to lower concentrations of sugar, like high-strain bees. Thus the effects of the ovary were directly linked to a key component of the pollen-hoarding syndrome—sugar sensitivity.

7.5.2 Vg Knockdown and Foraging Behavior

Another approach is to reduce vitellogenin circulating in the blood and look at effects on behavior. The prediction from the double-repressor and the reproductive-ground-plan models is that reduced Vg will result in an earlier onset of foraging because of earlier loss of circulating Vg and earlier increase in JH, and a nectar-foraging bias due to reduced Vg, as found in low-strain bees (Figure 7.5). Mindy Nelson and Kate Ihle tested this hypothesis by reducing the amount of circulating Vg in bees by RNA interference (Figure 7.6). Small pieces of double-stranded RNA (dsRNA) can inhibit the production of the protein encoded by the RNA. They injected *vg* dsRNA into the abdomen of newly emerged bees, the dsRNA was taken up by fat-body cells, and Vg production was greatly reduced. Control bees were injected with dsRNA for a gene that is not expressed in honey bees to test the effects of the injection minus the knockdown. Bees that had reduced Vg foraged earlier in life and demonstrated a bias for collecting nectar, as predicted.

7.5.3 Vg Knockdown and Sensitivity to Sugar

We were able to tie vitellogenin titers to foraging onset and foraging bias (specialization), but what about sensory sensitivity? Sucrose sensitivity

Fat Body

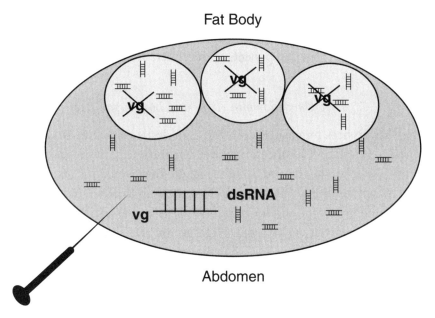

Figure 7.6. Cartoon of the effect of injecting double-stranded RNA (dsRNA) into the abdomen of a worker honey bee. RNA is normally found as a single strand after it is transcribed from the DNA in the nucleus of a cell. The messenger RNA migrates out of the nucleus and forms a template for protein synthesis in the cell, a process called translation. The dsRNA disrupts the translation process.

seems key to the phenotypic architecture of pollen hoarding and foraging behavior (Figure 5.14). Gro Amdam and Ricarda Scheiner injected newly emerged bees with *vg* dsRNA and tested them when they were 7 days old for their sucrose responses using the PER assay. Bees with reduced vitellogenin were more responsive to lower concentrations of sugar and water than were the control bees.

At first, this finding seemed contradictory to what we would have predicted because high-strain bees have higher titers of Vg and are more responsive throughout life. Vg knockdowns should have produced bees that were more like low-strain bees. However, Tanya Pankiw has shown that response to sugar increases with age and treatment with methoprene. Methoprene is an analog of JH that results in rapid behavioral development and earlier onset of foraging. It is likely that the age effect on sucrose response is modulated by JH, and JH is modulated by Vg.

Reduction of Vg increases JH and thus response to sugar. We were probably looking at an effect of Vg knockdown on the physiological aging of bees, one of the components of the pollen-hoarding syndrome.

7.5.4 Vg Knockdown in High- and Low-Strain Bees

High-strain bees forage earlier in life than do low-strain bees. In fact, Jennifer Fewell and I noticed very early in the selection program that many low-strain bees failed to forage at all. They would die after 40 to 50 days without initiating foraging. Gro Amdam knocked down Vg in newly emerged high- and low-strain workers and assayed their JH titers when they were 7 days old. High-strain bees, like wild types of other studies, had elevated JH, but the low-strain bees showed no response. Selection for low pollen hoarding apparently "uncoupled" the double-repressor regulatory system in low-strain bees but left it intact in highs.

Because the low-strain bees show no JH response to Vg, we had a tool to look at effects of the reproductive regulatory network by knocking down vitellogenin expression. If the reproductive regulatory network involving JH and Vg are regulating foraging behavior, then we should see a response to Vg knockdown in the high strain but not in the low. Kate Ihle used dsRNA to knock down Vg in newly emerged workers of the high- and low-pollen-hoarding strains. Knockdown high-strain bees foraged earlier in life and demonstrated a bias for nectar relative to the high-strain controls. This result was the same as that observed by Mindy Nelson and Kate Ihle discussed in Section 7.5.2. As predicted, there was no response to knockdown in the low-strain bees.

There is one necessary caveat to the story of the double-repressor relationship of Vg and JH. David Antonio and colleagues knocked down Vg in bees from a single colony of Africanized honey bees. The Vg-knockdown bees initiated long-term flights, assumed to be foraging flights, 3 to 4 days earlier in life than the controls. This was as we found. However, they found low levels of JH in 7-day-old knockdown bees. In this case, low Vg stimulated onset of early foraging even though JH stayed low. JH apparently played no role, as in the case of the allatectomized bees of Joe Sullivan (discussed in Section 7.4). It is clear that there is genetic variation for the way in which these networks are coupled, as demonstrated by differences between the high- and low-strain bees and colonies of European and Africanized bees from different studies. And there is still much to learn about the coupling of Vg and JH and their effects on foraging.

7.5.5 Ovary Transplants

Our research has shown a significant association between the number of ovarioles of workers and the age of onset of foraging and foraging bias for nectar and pollen. Our results were derived from correlative studies of high- and low-strain bees and wild-type bees. I wanted a more direct test of the causal relationship of ovary size and behavior. One idea was to be able to produce high-strain bees with fewer ovarioles and low-strain bees with more. We developed a diet based on modified commercially available royal jelly that we used to rear workers in petri dishes. The emerging workers had a wide range of ovariole numbers. My plan was to raise low-strain bees with big ovaries and see whether they behaved like high-strain bees. However, this idea was met with skepticism from my colleagues because they felt that the difference in feeding could cause many differences among bees that were correlated with the ovaries, so it would not be possible to construct a causal link from ovary to behavior. I abandoned that idea and turned to an idea posed by Gro Amdam, and together we tried ovary transplants.

Osman Kaftanoglu and my postdoctoral researcher Ying Wang began developing surgical procedures for removing ovaries from one bee and inserting them into the abdomen of another. Wang was trained in China both as a medical doctor and a molecular biologist. She has excellent surgical skills, at least on insects. They successfully developed a grafting technique where the recipient bees survived and the extra ovaries lived (Figure 7.7). In fact, the ovaries that were grafted into the abdomen responded to queenless conditions in the same way as the resident ovaries of the bees. We tested this by placing the bees with grafted

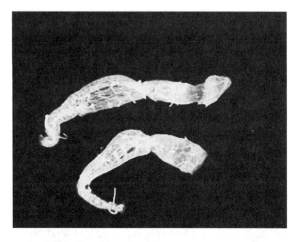

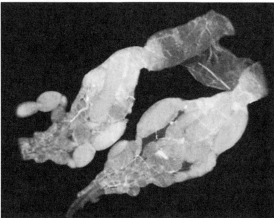

Figure 7.7. Ovaries from worker honey bees that were kept in a queenless colony for 5 (upper panel) and 18 (lower panel) days. Note that both ovaries of each pair appear healthy and are in the same stage of reproductive development. At 18 days, ovaries were activated and contained eggs in the transplanted and resident ovaries. Photo by Ying Wang.

ovaries in queenless, broodless hives and waiting for their ovaries to develop, as occurs normally. The resident ovaries become vitellogenic and contained developing eggs, but so did the transplanted ovaries, even though they were not connected to anything and were just floating in the hemolymph. They were indistinguishable from the resident ovaries. This showed that they could respond to signals that initiate reproductive activation. It is also likely that they were producing ecdysteroids and acting on the fat body, resulting in production of Vg. Colin Brent, a former postdoc in my lab, showed that worker honey bee ovaries produce ecdysteroids and that bigger ovaries produce more. This could be the mechanism linking ovaries to vitellogenin and behavior.

We then set up a test to see whether the bees with extra ovaries behaved differently. We implanted extra ovaries into 50 bees and immunologically inert glass beads of approximately the same weight into an equal number of control bees. The bees were placed in an observation hive, and my graduate student Adam Siegel watched them as they passed through their behavioral changes, culminating in the initiation of foraging. The bees with ovary implants made the within-hive behavioral transitions earlier and initiated foraging earlier in life than the controls, as expected. We were unable to do a foraging behavior study because the surgical procedure was too demanding to produce a sufficient number of bees. A sufficient statistical sample to distinguish between groups of foragers usually requires several hundred bees.

7.5.6 Verification of Genetic Architecture with Africanized Honey Bees

Adam Siegel used bees from the European and Africanized bee backcross discussed in Section 6.6 and looked at the relationship between foraging behavior and the number of ovarioles. He found a significant relationship between the number of ovarioles and both the amount of nectar collected and the sugar concentration of the nectar. In other words, the size of the ovary was affecting the perception of the nectar reward. This was a significant result because the bees for the mapping study and the bees for the behavioral study came from the same sources, and these sources were not derived from the high and low strains. Hence he independently confirmed the QTL, ovary, and behavior relationships found in the selected strains.

7.6 Difficulties with the Vitellogenin Foraging Model

Like most models in biology, ours cannot explain all the observed results. That is why I call it a heuristic model. It serves as a framework on
which to hang data and thoughts and for the generation of new questions. Ours is certainly false, but where and how? First, Vg and JH must
be working together. Joseph Sullivan and colleagues showed that JH is
not necessary for the onset of foraging by removing the JH-producing
corpora allata from bees—they foraged anyway. We assume that because JH inhibits Vg production, the allatectomized bees had higher
titers of Vg but foraged at about the same time as the undisturbed controls. There must be a JH-independent pathway, but there also must be
a Vg-independent pathway. We showed that knocking down Vg in low-
strain bees does not increase JH, but the knockdown bees still forage at
the same age as the controls, in which case neither Vg nor JH has an effect. The ways in which JH and Vg interact to affect foraging onset need
to be resolved.

High-strain bees have higher levels of Vg in the blood over the first
10 days of adult life. The double-repressor hypothesis, in its original
form, suggested that blood titers in nurse bees decrease as they feed
brood. The fat-body cells are still producing Vg, but the rate of production is less than the rate of transfer of Vg through the brood food
glands to developing larvae. High-strain bees forage earlier in life, so
their blood titers of Vg must fall very rapidly compared with low-
strain bees. Therefore, we can assume that high-strain bees feed more
larvae.

Adam Siegel looked at the nurse behavior of high-strain bees. He
uniquely marked 600 high-strain and 600 low-strain bees and placed
300 of each in two observation hives. For behavioral studies, we often
tag workers with plastic disks that are numbered. The disks are glued to
the thoraces of bees and usually remain throughout life. Observation
hives are also a common tool for behavioral research. They typically
consist of two to four frames with combs stacked vertically and have
glass sides so that the activities of the bees can be observed (Figure 7.8).
Siegel observed the marked cohorts for about 30 days, through the nor-

Figure 7.8. Observation bee hive. Photo by Kathy Keatley Garvey.

mal transitions of nest cleaning, brood care, nest construction and maintenance, food processing, guarding, and foraging. High-strain bees initiated and ceased brood care (including feeding larvae) at younger ages than the low-strain workers but did not engage in more brood care. High- and low-strain bees did not differ in food exchange between adults. Therefore, we cannot explain the early and rapid decline in Vg titers in high-strain bees on the basis of differential transfer of Vg to brood and adults.

7.7 Summary Comments

The reproductive-ground-plan hypothesis provides the needed mechanism to integrate the phenotypic and genetic architectures of the pollen-hoarding syndrome. It simplifies our understanding. It proposes that an ancient reproductive regulatory network of insects has been used and remodeled by natural selection into an adaptive division of labor. Evidence in support of this hypothesis comes from studies of behavior, anatomy and physiology, gene silencing, hormone analysis, organ transplants, and QTL mapping.

For more than 20 years, we selected for a single trait, the amount of pollen stored in the comb. It resulted in changes in the high- and low-pollen-hoarding strains at levels of the gene, most likely in some genes involved in insulin-insulin-like signaling pathways, anatomical features of the ovary, hormonal interactions, neurophysiology, sensory-response systems, and behavior. But one result suggests that we now look at development: the number of ovarioles correlates with gene expression, sensory responses, and behavior (Figure 5.14). The number of ovarioles in bees is determined during larval development. So that is where we looked.

Suggested Reading

Amdam, G. V., Csondes, A., Fondrk, M. K., and Page, R. E. 2006. Complex social behaviour derived from maternal reproductive traits. *Nature* 439:76–78.

Amdam, G. V., Ihle, K. E., and Page, R. E. 2009. Regulation of honeybee worker (*Apis mellifera*) life histories by vitellogenin. In *Hormones, Brain and Behavior*, 2nd ed., vol. 2, ed. D. W. Pfaff, A. P. Arnold, A. M. Etgen, S. E. Fahrbach, et al. San Diego: Academic Press, pp. 1003–1025.

Amdam, G. V., Nilsen, K. A., Norberg, K., Fondrk, M. K., et al. 2007. Variation in endocrine signaling underlies variation in social life history. *Am. Nat.* 170:37–46.

Amdam, G. V., Norberg, N., Fondrk, M. K., and Page, R. E. 2004. Reproductive ground plan may mediate colony-level effects on individual foraging behavior in honey bees. *Proc. Natl. Acad. Sci. USA* 101:11350–11355.

Amdam, G. V., Norberg, K., Hagen, A., and Omholt, S. W. 2003. Social exploitation of vitellogenin. *Proc. Natl. Acad. Sci. USA* 100:1799–1802.

Amdam, G. V., and Omholt, S. W. 2002. The regulatory anatomy of honeybee lifespan. *J. Theor. Biol.* 216:209–228.

Amdam, G. V., and Omholt, S. W. 2003. The hive bee to forager transition in honey bee colonies: The double repressor hypothesis. *J. Theor. Biol.* 223:451–464.

Amdam, G. V., and Page, R. E. 2009. Oldroyd and Beekman do not test ground plan hypothesis that explains origins of social behavior. *PLoS Biol.* 6:e56r2248.

Amdam, G. V., and Page, R. E. 2010. The developmental genetics and physiology of honeybee societies. *Animal Behavior* 79:973–980.

Amdam, G. V., Page, R. E., Fondrk, M. K., and Brent, C. S. 2010. Hormone response to bidirectional selection on social behavior. *Evol. Devel.* 12:428–436.

Amdam, G. V., Rueppell, O., Fondrk, M. K., Page, R. E., et al. 2009. The nurse's load: Early-life exposure to brood-rearing affects behavior and lifespan in honey bees. *Exp. Gerontol.* 44:467–471.

Antonio, D. S. M., Guidugli-Lazzarini, K. R., do Nascimento, A. M., Simões, Z. L. P., et al. 2008. RNAi-mediated silencing of vitellogenin gene function turns honeybee *(Apis mellifera)* workers into extremely precocious foragers. *Naturwissenschaften* 95:953–961.

Graham, A. M., Munday, M. D., Kaftanoglu, O., Page, R. E., et al. 2011. Support for the reproductive ground plan hypothesis of social evolution and major QTL for ovary traits of Africanized worker honey bees *(Apis mellifera* L.). *BMC Evol. Biol.* 11:95. doi:10.1186/1471-2148-11-95.

Nelson, C. M., Ihle, K. E., Fondrk, M. K., Page, R. E., et al. 2007. The gene vitellogenin has multiple coordinating effects on social organization. *PloS Biol.* 5:673–677.

Oldroyd, B. P., and Beekman, M. 2008. Effects of selection for honeybee worker reproduction on foraging traits. *PLoS Biol.* 6:e56. doi:10.1371/journal.pbio .0060056.

Page, R. E., Scheiner, R., Erber, J., and Amdam, G. V. 2006. The development and evolution of division of labor and foraging specialization in a social insect. *Curr. Topics in Devel. Biol.* 74:253–286.

Rueppell, O., Metheny, J. D., Linksvayer, T., Fondrk, M. K., et al. 2011. Genetic architecture of ovary size and asymmetry in European honeybee workers. *Heredity* 106:894–903.

Siegel, A. J., Kaftanoglu, O., Fondrk, M. K., Smith, N. R., et al. 2012. Ovarian regulation of foraging division of labour in Africanised honey bees. *Anim. Behav.* 83:653–658.

Sullivan, J. P., Fahrbach, S. E., Harrison, J. F., Capaldi, E. A., et al. 2003. Juvenile hormone and division of labor in honey bee colonies: Effects of allatectomy on flight behavior and metabolism. *J. Exp. Biol.* 206:2287–2296.

Sullivan, J. P., Jassim, O., Fahrbach, S. E., and Robinson, G. E. 2000. Juvenile hormone paces behavioral development in the adult worker honey bee. *Horm. Behav.* 37:1–14.

Thompson, G. J., Yockey, H., Lim, J., and Oldroyd, B. P. 2007. Experimental manipulation of ovary activation and gene expression in honey bee *(Apis mellifera)* queens and workers: Testing hypotheses of reproductive regulation. *J. Exp. Zool.* 307A:600–610.

Wang, Y., Amdam, G. V., Wallrichs, M. A., Fondrk, M. K., et al. 2009. *PDK1* and *HR46* gene homologs tie social behavior to ovary signals. *PLoS ONE* 4:e4899. doi:10.1371/journal.pone.0004899.

Wang, Y., Kaftanoglu, O., Siegel, A., Page, R. E., et al. 2010. Surgically increased ovarian mass in the honey bee confirms link between reproductive physiology and worker behavior. *J. Insect Physiol.* 56:1816–1824.

West-Eberhard, M. J. 1987. Flexible strategy and social evolution. In *Animal Societies: Theories and Fact,* ed. Y. Itō, J. L. Brown, and J. Kikkawa. Tokyo: Japan Science Society Press, pp. 35–51.

West-Eberhard, M. J. 1996. Wasp societies as microcosms for the study of development and evolution. In *Natural History and Evolution of Paper-Wasps,* ed. S. Turillazzi and M. J. West-Eberhard. New York: Oxford University Press, pp. 290–317.

— 8 —

Developmental Regulation
of Reproduction

In Chapter 7, I hypothesized that the reproductive signaling networks had been used by natural selection to build foraging division of labor in honey bees. I proposed that foraging for pollen (protein) and nectar (carbohydrate) reflected behavioral responses linked to the developmental states of ovaries. The fate of the ovary is determined during larval development. We need to look there for the signatures of pollen-hoarding selection.

In the following sections, I first describe the queen and worker phenotypes and then show how they are determined by nutritional inputs provided to the larvae by the nurse bees. I make the argument that the developmental program of a honey bee larva is shared with the nutritional program presented by the nurses, and larvae and nurses coadapt to produce the appropriate phenotypes. They share a joint social genotype. Then I dissect the nutritional programs of queen and worker larvae on the basis of food quantity, quality, and temporal presentation and show the effects on the development of queen and worker traits. I locate the critical stage during worker development where nurse bees can manipulate the ovaries of their sister larvae and show how pollen-hoarding selection has apparently affected larval development and thereby the foraging behavior of bees.

8.1 Queen and Worker Phenotypes

The most obvious evidence for higher levels of selection operating on honey bee colonies is the difference in the phenotypes of the queen and the workers. Queens and workers represent different castes of females. The queen is highly reproductive, while workers are facultatively sterile. Workers normally do not produce eggs except in the absence of the queen and larvae. Workers and queens can develop from the same eggs. Every fertilized egg is capable of producing either a queen or a worker. Workers are reared in the many thousands of cells composing the combs of a nest. Queens are raised in special, larger cells normally located on the lower edges of combs near the center of the nest. Beekeepers take advantage of this by removing young larvae from worker-sized comb cells and putting them in queen-sized cells. This technique is called grafting (Figure 8.1). Beekeepers can raise dozens of new queens in a single colony by manipulating it to induce a queen-rearing stimulus and then placing

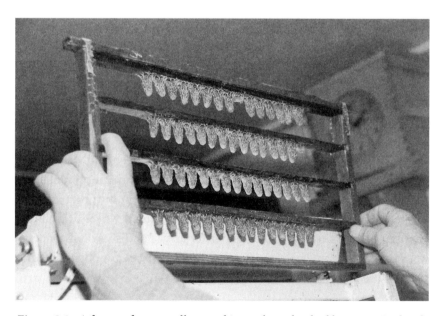

Figure 8.1. A frame of queen cells reared in a colony that had been manipulated to induce queen-rearing behavior. Photo by Kathy Keatley Garvey.

larvae grafted from worker-sized cells in artificially produced queen-sized cells. Grafted larvae need to be less than about 2 to 2½ days old to develop into queens. The resulting queens are normally indistinguishable from queens raised by colonies under normal queen-rearing conditions. Compared with workers, queens are about twice the size and weight (about 200 milligrams), have a smooth sting rather than a barbed one, have a shorter tongue, do not have a pollen-carrying apparatus on their hind legs, and have notched mandibles, larger mandibular glands, no wax glands, a fully developed spermatheca (a spherical organ for storing sperm), faster development time from egg to adult (16 versus 21 days), and on average around 140 to 160 ovarioles (egg-making filaments of the ovary) per ovary, compared with 2 to 6 for workers (Figure 8.2).

The quality (sugar content) and quantity of food fed to larvae by nurse bees determine the fate of the developing larva: queen or worker.

Figure 8.2. A queen with a retinue of workers. The retinue constantly feeds and grooms the queen. Photo by Harry Laidlaw.

Queen and worker larvae are fed proteinaceous secretions from hypopharyngeal glands in the heads of the nurse bees. Food fed to queens is often called royal jelly, while food fed to workers is called worker jelly. Most of the proteins produced by the hypopharyngeal glands of nurse bees belong to a family called major royal jelly proteins. The glandular food appears to be the same for worker jelly and royal jelly, except that more sugar is added to the queen food initially.

Here it is important to note that for decades apicultural researchers have led a quest to find a special substance that is added only to the food of queens that is responsible for "unlocking" a developmental program or throws a developmental switch that results in the production of the queen phenotype. This quest has had many twists and turns. The most recent special substance, royalactin, was proposed by Masaki Kamukura to be one of the major royal jelly proteins found in brood food. In an elegant study, he showed that royalactin is necessary to make a queen. However, although it may be necessary for developing a queen, it does not qualitatively differentiate queen from worker development because it is fed in the same proportion to developing worker larvae. Today, after at least 60 years of the quest, the only consistent and independently confirmed result is that the food of worker and queen larvae differs in the amount of sugar it contains.

Gene Robinson and I, along with his student Naomi Arensen, studied the feeding behavior of individual nurse bees when they had a choice to feed worker or queen larvae. We found that individuals switched readily between the two larval types in successive feeding bouts. I find it difficult to believe that individual nurse bees, with thousands of larvae to feed, differentially control the composition of the glandular secretions they feed to queen and worker larvae in successive feeding bouts. It is easy to believe that nurse bees differentially add nectar or honey (sources of sugar) to the food of queen and worker larvae.

Sugar acts as a phagostimulant, causing the larvae to eat more, but it also probably increases metabolism and the production of juvenile hormone (JH), a growth-regulating hormone found in all insects. During the first two days of larval development, the queen larvae respond to the extra sugar in their food with elevated respiration but do not grow faster than worker larvae. In fact, by day three, the worker larvae are larger than queen larvae, probably because they both have access to the same quantities of food, but queens have higher metabolism. JH titers are elevated in queen larvae relative to worker larvae throughout larval development. Sugar concentration goes up in worker food after the fourth instar; it is necessary for them to pupate. (An instar is a developmental stage that is marked by a moult, the shedding and regrowing of the external skeleton of insects. Honey bee larvae moult roughly every 24 hours. Therefore, there is a rough correspondence between the age of larvae and the instar [Figure 8.3].) Nurses deposit a large excess of food in queen cells before they are capped on the fifth day (fifth instar) after hatching,

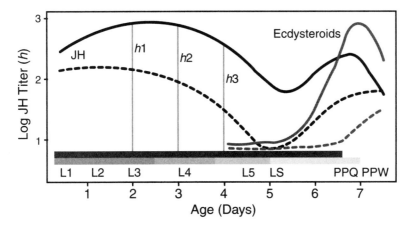

Figure 8.3. Feeding regimes and hormonal profiles of developing queens and workers. The JH titers of queens (solid black line) and workers (dashed black line) respond to the feeding regimes imposed by nurses. Ecdysteroid titers are shown by gray lines. Queen food is unrestricted and contains about 12 percent sugar (top bar), whereas worker food changes over development (multishaded bar). During the first three instars (L1–L3) worker food is unrestricted but contains only around 4 percent sugar. Feeding is restricted in the fourth instar, and in the fifth the sugar content is increased. After nurses seal the worker cells (LS), workers starve through to the prepupal stage (PPW), whereas queen cells are mass provisioned at sealing, so queens continue feeding until the prepupal stage (PPQ). Ecdysteroid levels rise after the cells are sealed, signaling the moult into the pupal stage. From Leimar et al. 2012. "Development and evolution of caste dimorphism in honeybees—a modeling approach," *Ecology and Evolution* 2:3098–3109.

and the queens continue to feed and gain weight for an additional two days. Worker larvae are capped on the sixth day without surplus food and lose weight when they spin their cocoon before pupation.

Until the fourth instar, queens and workers both have a full complement of ovariole primordia (clusters of cells that develop into ovarioles in the pupal stage), enough to make 300 or more ovarioles. Elevated titers of JH in queens probably initiate changes in target tissues and prime specific tissues for worker-queen differentiation in response to ecdysteroids, important growth hormones of insects. In addition, JH protects the ovarioles from programmed cell death that begins in the fourth and fifth larval instars, as shown by Ines Schmidt Capella and Klaus Hartfelder. Programmed cell death is a normal part of developmental processes where cells that are no longer needed, or are needed for structure but do not need to be living cells, undergo programmed death. In queens, most of the 300 or so ovarioles are rescued, while most of them fail in workers. In the mid- to late fifth instar, an increase of JH occurs in both workers and queens, resulting in an increase in ecdysteroid hormones, and initiates metamorphosis and differentiation between the worker and queen phenotypes.

It appears that the queen phenotype is the default. It is the easiest for the workers to achieve when one considers the requirements. All that is needed is to provide a large cell with food in excess and add sugar. The sugar in royal jelly is composed of fructose, glucose, and a small amount of sucrose. Fructose is the majority sugar and is what is found in most nectar and honey. It makes sense that development would be sensitive to this nutritional component that is central to honey bee survival. To make a worker requires the orchestration of a complex feeding program: reduce sugar, feed unlimited food for two days and then regulate (restrict) it, increase sugar content of food in the fourth instar, and starve for the final day before moulting.

8.2 Nurses and Larvae Share Developmental Programs

Larval development is jointly controlled by the physiology and behavior of the nurse bees and the developmental regulatory programs of the larvae. They evolved in concert to produce the adult phenotype. Larvae produce a blend of carbon compounds called brood pheromone that influences foraging behavior and the feeding behavior of nurses. It stimulates the development of the hypopharyngeal glands (brood food glands) of nurse bees, and it delays the onset of foraging in young bees, so more bees perform preforaging tasks, such as nurse behavior. In colonies treated with supplemental synthetic brood pheromone, it increases pollen foraging, but not the amount of pollen stored, so there is more pollen consumption and more protein available in the brood food glands of nurse bees. The result is an increase in the number of larvae reared by the colony. Larvae provide signals to the nurses that they are present in the cell and are hungry. Larvae also respond to the nutritional delivery program provided by nurse bees, convert the nutritional contributions of the nurse bees into tissues, and thereby may control the investment in ovaries versus other body tissues. Therefore, larvae affect their development through their genetic-developmental program and their effects on the nurse bees that feed them. We can think of this as the "larval module" of development (Figure 8.4).

Nurse bees have evolved a developmental program that includes their responses to the pheromonal signals and other cues derived from the developing larvae, as well as the sizes of the cell in which the larvae develop, that result in the quality, quantity, and temporal delivery of food. The responses of the larvae are matched to the program of the nurses. For example, the nurses feed larvae ad lib during the first 2 days. Ad lib feeding during the first 2 days is probably a mechanism to prevent starvation and death of the tiny larvae. During this time they need a tremendous growth rate in order to reach the needed size and weight within the 6-day window of larval development. They die rapidly when they run out of food. Older larvae are much more resistant to starvation (unpublished data). Larvae may provide pheromonal signals to inform nurse bees when they consume the food in their cells, stimulating the nurses to replenish. The nurses feed the worker larvae a diet low in sugar, but the larvae must respond to lower sugar, perhaps by eating less, and develop like a worker rather than a queen. After 2 days, workers receive restricted food from nurses but may increase their

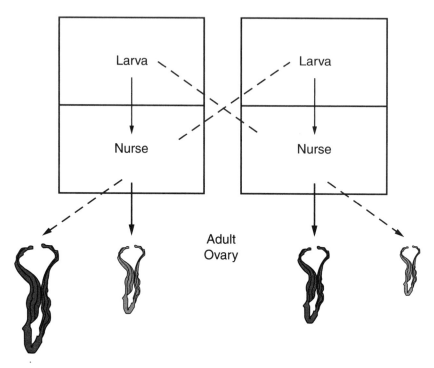

Figure 8.4. Illustration of the effects of the larval and nurse modules on the resultant ovary-size phenotype. The left box shows the combination of the low-strain larval development module and nurse bee modules that were selected together during the pollen-hoarding selection program. The genes that were selected and responsible for the larval and nurse modules constitute a social genotype. The right box shows the high strain. High- and low-strain bees were cross-fostered. The resulting phenotypes are shown at the bottom. On the left are the ovary phenotypes of high- and low-strain workers that were raised by low-strain nurses; on the right, those of workers raised by high-strain nurses. The nurse modules interact with the larval development modules such that high-strain larvae raised by low-strain nurses end up with more ovarioles than those raised by nurses from their own colonies.

hunger signal, and the nurses correspondingly may become more or less sensitive to it. One can think of the nurse bee feeding program and responses to larval feeding signals and cues as the "nurse bee developmental module." These modules interact to produce a worker (or queen) phenotype. One can think of the combination of genes affecting the coevolved larval and nurse bee developmental modules as a social genotype (Figure 8.4).

8.2.1 In Vitro Rearing Studies

High- and low-strain bees vary in the age of foraging onset and foraging behavior at least in part because they have different-sized ovaries (see Chapter 7). I wanted to rear high- and low-strain bees in the laboratory (in vitro) and vary their nutrition in order to make high-strain workers with smaller ovaries, the size of low-strain bees, and low-strain workers with larger ovaries. I thought that I would then look at their foraging behavior. (We never did the experiments.) So Osman Kaftanoglu developed diets, derived from frozen royal jelly, that contained differing amounts of added sugar, protein, and water and fed different amounts with different food delivery schedules to larvae from high- and low-strain bees, commercial wild-type European bees, and Africanized bees. This was similar to an ensemble modeling approach (Chapter 2). We were exploring the total possible phenotypic space that could be generated by varying nutritional and genetic parameters.

Tim Linksvayer and Kaftanoglu hand-raised about 2,500 individuals. In addition, workers and queens were reared in their own colonies by their own nurse bees for comparison. They weighed the adults and dissected the ovaries to count the number of ovarioles. They also scored the adults for other characters that differ between queens and workers: pollen baskets, spermathecae, and notched mandibles. The feeding studies revealed a strong general relationship between body mass and ovariole number, and that the phenotypes of queens and workers were not discretely determined by the developmental programs of the larva. In other words, we uncovered a cloud of phenotypes that are never seen when nurse bees rear queens and workers (Figure 8.5). The other characters also varied from workerlike to queenlike with all intermediate stages, as previously shown by Selim Dedej and his colleagues. Workers and queens raised by nurse bees occupied a much smaller part of the total phenotypic space. The nurse bees constrained the development of the larvae.

8.2.2 Cross-Fostering Experiments

To illustrate the joint operation of the larval and nurse bee modules, we conducted cross-fostering experiments. These are similar to

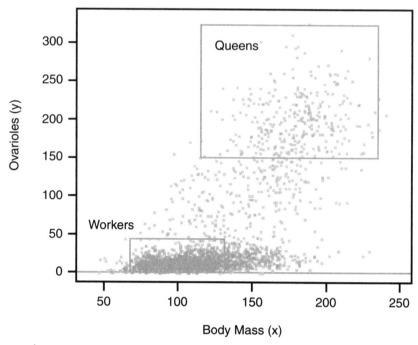

Figure 8.5. Ovariole number plotted against body mass (wet weight) for 3,610 individuals reared in the laboratory (in vitro) or in their own colonies. Boxes define the total phenotypic space of workers and queens when they are reared in colonies by nurse bees. All points outside the boxes represent phenotypes that are possible but are not produced in colonies. Bees reared in vitro covered the entire phenotypic space. From Leimar et al. 2012. "Development and evolution of caste dimorphism in honeybees—a modeling approach." *Ecology and Evolution* 2:3098–3109.

common-garden experiments (see Figure 4.5). We raised high- and low-strain bees from eggs together in high- and low-strain colonies. Therefore, high- and low-strain larvae were raised by high- and low-strain nurse bees. We looked at body mass (wet weight, the total weight including water) and ovariole number because both varied between the high- and low-strain bees. High-strain bees on average had lower body mass and more ovarioles. We could separate the effects of the nurse bee environment from the genotype of the larva. High-strain larvae had more ovarioles than low-strain larvae regardless of the nurse bees raising them. However, when high-strain workers were raised in low-strain colonies by low-strain nurses, they had more ovarioles than when they were raised by nurses of their own strain. This demonstrates that the nurse module for development affected ovary development and differed between the high and low strains (Figure 8.4).

8.2.3 Starvation Experiments

Because nutrition affects JH titers throughout larval development, and the JH titers during the fourth and early fifth instars are critical for rescuing ovarioles from programmed cell death, we starved early fifth-instar larvae. We were interested in the nutritional sensitivity of larvae during different stages of development in order to better understand how nurse bees might control development. Ying Wang and Osman Kaftanoglu (unpublished data) removed combs with larvae from colonies and placed them in an incubator for 6 hours. During this time they were not fed. Control larvae continued to be fed by nurse bees. The frames with the larvae were then returned to the hive, where they completed development. The 6 hours of "starvation" (6 hours out of a total of 6 days of larval development) had a profound effect on ovary development. Control bees had an average of about nine ovarioles while starved bees had an average of about five, a reduction of nearly 50 percent. This suggests that the timing of food delivery to developing larvae at this stage may be a mechanism whereby the nurse bees control ovary development in developing larvae, part of the nurse bee module (Figure 8.4).

8.3 Developmental Signatures of Colony-Level Artificial Selection

Worker honey bees are responsible for the differences in the pollen-hoarding phenotypes between the high and low strains. Individual bees develop from egg to adult. Even newly emerged adults show differences in sensorimotor-system responses and in some key neurobiochemical, insulin-insulin-like signaling pathways, reproductive hormones, and signaling proteins and peptides. They also show differences in ovary anatomy. Therefore, there must have been fundamental differences during development. We found several developmental signatures of our selection program. A signature is a developmental difference that leads to a definable difference in phenotype that affects the trait we were selecting, pollen hoarding.

8.3.1 Ovariole Number and Elevated JH
Titers in Developing Larvae

High-strain bees have more ovarioles than do low-strain bees, probably because they have higher JH titers during larval development, which protect more ovarioles in high-strain bees from programmed cell death. Colin Brent and Gro Amdam sampled developing high- and low-strain larvae at 3, 4, and 5 days after hatching and tested their hemolymph for JH. High-strain bees had higher titers on all three days, the critical window where ovarioles are "rescued" from cell death.

8.3.2 Sensitivity of Ovary Size to Body Size

High-strain bees show more ovary response (increase in ovarioles) with increasing body mass across the phenotypic range of normal worker body mass (Figure 8.6). This shows that the growth relationship is not fixed and is selectable. Pollen-hoarding selection steepened the developmental relationship in high-strain relative to low-strain bees, resulting in more ovarioles.

8.3.3 Uncoupling of the JH/Vg Switch

The hypothesis of a JH/Vg double-repressor switch has been confirmed repeatedly for honey bees and is a robust finding (Section 7.5.4). It is interesting that it seems to be unique to honey bees, at least in species looked at so far. Usually JH regulates Vg; Vg does not regulate JH. So we can look at the JH/Vg double-repressor switch as a signa-

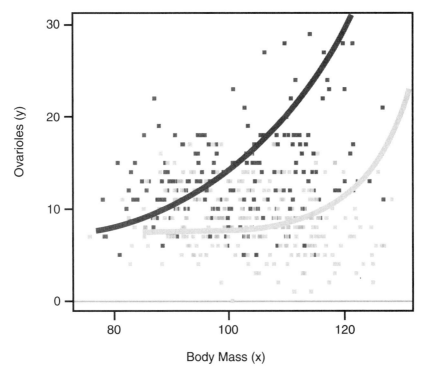

Figure 8.6. Relationships of body mass to ovarioles of high-strain (dark points) and low-strain (light points) bees that were raised in vitro and by nurse bees (see Section 8.2.1). From Leimar et al. 2012. "Development and evolution of caste dimorphism in honeybees—a modeling approach." *Ecology and Evolution* 2:3098–3109.

ture of natural selection, affecting foraging onset and division of labor. We further modified it through our selection for pollen hoarding. In particular, selection for low pollen hoarding uncoupled the joint control mechanism of JH and Vg (see Section 7.5.4).

8.4 Summary Comments

I have traced the extended reach of colony-level selection through many levels of biological organization and have shown the network of interactions within and between levels that changed as a consequence. It is a complex web with integrating control systems, such as the reproductive regulatory network. At times we have been on our hands and knees looking at grains of sand one at a time. At other times, we have climbed up a few hundred meters to get a better view of the beach. In the next chapter, we will go to 10,000 meters, and I will give a larger overview of the regulatory architecture of pollen hoarding from the genotype to the social phenotype.

Suggested Reading

Alexander, B., and Rozen, J. G. 1987. Ovaries, ovarioles and oocytes in parasitic bees (Hymenoptera: Apoidea). *Pan-Pac. Entomol.* 63:155–164.

Allsopp, M. H., Calis, J. N. M., and Boot, W. J. 2003. Differential feeding of worker larvae affects caste characters in the Cape honeybee, *Apis mellifera capensis*. *Behav. Ecol. Sociobiol.* 54:555–561.

Amdam, G. V., Page, R. E., Fondrk, M. K., and Brent, C. S. 2010. Hormone response to bidirectional selection on social behavior. *Evol. Devel.* 12:428–436.

Antonialli, W. F., and da Cruz-Landim, C. 2009. Effect of topic application of juvenile hormone on the ovarian development of worker larvae of *Apis mellifera* Linnaeus (Hymenoptera, Apidae). *Rev. Bras. Entomol.* 53:115–120.

Asencot, M., and Lensky, Y. 1976. Effect of sugars and juvenile hormone on differentiation of female honeybee larvae (*Apis mellifera* L.) to queens. *Life Sci.* 18:693–700.

Asencot, M., and Lensky, Y. 1984. Juvenile hormone induction of "queenliness" on female honey bee (*Apis mellifera* L.) larvae reared on worker jelly and on stored royal jelly. *Comp. Biochem. Physiol. B—Biochem. Mol. Biol.* 78:109–117.

Asencot, M., and Lensky, Y. 1985. The phagostimulatory effect of sugars on the induction of queenliness in female honeybee (*Apis mallifera*) larvae. *Comp. Biochem. Physiol.* 81A:203–208.

Asencot, M., and Lensky, Y. 1988. The effect of soluble sugars in stored royal jelly on the differentiation of female honeybee (*Apis mellifera*) larvae to queens. *Insect Biochem.* 18:127–133.

Calis, N. M., Boot, W. J., Allsopp, M. H., and Beekman, M. 2002. Getting more than a fair share: Nutrition of worker larvae related to social parasitism in the cape honey bee *Apis mellifera capensis*. *Apidologie* 33:193–202.

Dedej, S., Hartfelder, K., Aumeier, P., Rosenkranz, P., et al. 1998. Caste determination is a sequential process: Effect of larval age at grafting on ovariole number, hind leg size and cephalic volatiles in the honey bee *(Apis mellifera carnica). J. Apic. Res.* 37:183–190.

Dietz, A., and Haydak, M. H. 1971. Caste determination in honey bees. I. The significance of moisture in larval food. *J. Exp. Zool.* 177:353–358.

Dietz, A., Hermann, H. R., and Blum, M. S. 1979. The role of exogenous JH1, JHIII and anti-JH (precocene II) on queen induction of 4.5-day-old worker honey bee larvae. *J. Insect Physiol.* 25:505–512.

Eder, J., Kremer, J. P., and Rembold, H. 1983. Correlation of cytochrome c titer and respiration in *Apis mellifera*—Adaptive response to caste differentiation defines workers, intercastes and queens. *Comp. Biochem. Physiol. B— Biochem. Mol. Biol.* 76:703–716.

Graham, A. M., Munday, M. D., Kaftanoglu, O., Page, R. E., et al. 2011. Support for the reproductive ground plan hypothesis of social evolution and major QTL for ovary traits of Africanized worker honey bees *(Apis mellifera* L.). *BMC Evol. Biol.* 11:95. doi:10.1186/1471-2148-11-95.

Hartfelder, K., and Engels, W. 1998. Social insect polymorphism: Hormonal regulation of plasticity in development and reproduction in the honey bee. *Curr. Top. Dev. Biol.* 40:45–77.

Hartfelder, K., and Steinbrück, G. 1997. Germ cell cluster formation and cell death are alternatives in caste-specific differentiation of the larval honey bee ovary. *Invertebr. Reprod. Dev.* 31:237–250.

Haydak, M. H. 1943. Larval food and development of castes in the honeybee. *J. Econ. Entomol.* 36:778–792.

Honek, A. 1993. Intraspecific variation in body size and fecundity in insects: A general relationship. *Oikos* 66:483–492.

Huang, Z. Y., and Otis, G. W. 1991. Inspection and feeding of larvae by worker honey bees (Hymenoptera: Apidae): Effect of starvation and food quality. *J. Insect Behav.* 4:305–317.

Kamakura, M. 2011. Royalactin induces queen differentiation in honeybees. *Nature* 473:478–483.

Laidlaw, H. H., and Page, R. E. 1997. *Queen Rearing and Bee Breeding.* Cheshire, CT: Wicwas Press.

Leimar, O., Hartfelder, K., Laubichler, M. D., and Page, R. E. 2012. Development and evolution of caste dimorphism in honeybees—a modeling approach. *Ecol. Evol.* 2:3098–3109.

Linksvayer, T. A., Kaftanoglu, O., Akyol, E., Blatch, S., et al. 2011. Larval and nurse worker control of developmental plasticity and the evolution of honey bee queen-worker dimorphism. *J. Evol. Biol.* 24:1939–1948.

Martins, G. F., and Serraão, J. E. 2004. Comparative study of the ovaries in some Brazilian bees (Hymenoptera; Apoidea). *Papéis Avulsos de Zool., S. Paulo* 44:45–53.

Osborne, K. E., and Oldroyd, B. P. 1999. Possible causes of reproductive dominance during emergency queen rearing by honeybees. *Anim. Behav.* 58:267–272.

Pankiw, T., Page, R. E., and Fondrk, M. K. 1998. Brood pheromone stimulates pollen foraging in honey bees *(Apis mellifera). Behav. Ecol. Sociobiol.* 44:193–198.

Rachinsky, A., and Engels, W. 1995. Caste development in honeybees *(Apis mellifera):* Juvenile hormone turns on ecdysteroids. *Naturwissenschaften* 82:378–379.

Rachinsky, A., and Hartfelder, K. 1990. Corpora allata activity, a prime regulating element for caste-specific juvenile hormone titre in honey bee larvae *(Apis mellifera carnica). J. Insect Physiol.* 3:189–194.

Rachinsky, A., Strambi, C., Strambi, A., and Hartfelder, K. 1990. Caste and metamorphosis—Hemolymph titers of juvenile hormone and ecdysteroids in last instar honey bee larvae. *Gen. Comp. Endocrinol.* 79:31–38.

Rembold, H. 1987. Caste-specific modulation of juvenile hormone titers in *Apis mellifera. Insect Biochem.* 17:1003–1006.

Rembold, H., Kremer, J. P., and Ulrich, G. M. 1980. Characterization of postembryonic developmental stages of the female castes of the honey bee, *Apis mellifera* L. *Apidologie* 11:29–38.

Richards, K. W. 1994. Ovarian development, ovariole number, and relationship to body size in *Psithyrus* spp. (Hymenoptera: Apidae) in southern Alberta. *J. Kansas Entomol. Soc.* 67:156–168.

Robinson, G. E., Page, R. E., and Arensen, N. 1994. Genotypic differences in brood rearing in honey bee colonies: Context specific? *Behav. Ecol. Sociobiol.* 34:125–137.

Sagili, R. R., and Pankiw, T. 2009. Effect of brood pheromone modulated brood rearing behaviors on honey bee *(Apis mellifera* L.) colony growth. *J. Insect Behav.* 22:239–249.

Schmidt Capella, I. C., and Hartfelder, K. 1998. Juvenile hormone effect on DNA synthesis and apoptosis in caste-specific differentiation of the larval honey bee *(Apis mellifera* L.) ovary. *J. Insect Physiol.* 44:385–391.

Schmidt Capella, I. C., and Hartfelder, K. 2002. Juvenile-hormone-dependent interaction of actin and spectrin is crucial for polymorphic differentiation of the larval honey bee ovary. *Cell Tissue Res.* 307:265–272.

Schmitzová, J., Klaudiny, J., Albert, S., Schröder, W., et al. 1998. A family of major royal jelly proteins of the honeybee *Apis mellifera* L. *Cell. Mol. Life Sci.* 54:1020–1030.

Shuel, R. W., and Dixon, S. E. 1959. Studies in the mode of action of royal jelly in honeybee development. II. Respiration of newly-emerged larvae on various substrates. *Canadian J. Zool.* 37:803–813.

Shuel, R. W., and Dixon, S. E. 1960. The early establishment of dimorphism in the female honey bee, *Apis mellifera. Insect Soc.* 7:265–282.

Shuel, R. W., and Dixon, S. E. 1968. The importance of sugar for the pupation of the worker honeybee. *J. Apic. Res.* 7:109–112.

Stabe, H. 1930. The rate of growth of worker, drone and queen larvae of the honeybee, *Apis mellifera* Linn. *J. Econ. Entomol.* 23:447–453.

Tsuruda, J. M., and Page, R. E. 2009. The effects of young brood on the foraging behavior of two strains of honeybees *(Apis mellifera). Behav. Ecol. Sociobiol.* 64:161–167.

Weaver, N. 1957. Effects of larval age on dimorphic differentiation of the female honey bee. *Ann. Entomol. Soc. Am.* 50:283–294.

Weiss, K. 1974. Zur Frage des Königinnengewichtes in Abhängigkeit von Umlarvalter und Larvenversorgung. *Apidologie* 5:127–147.

Wirtz, P., and Beetsma, J. 1972. Short communication: Induction of caste differentiation in the honeybee *(Apis mellifera)* by juvenile hormone. *Ent. Exp. Appl.* 15:517–520.

— 9 —

The Regulatory Architecture
of Pollen Hoarding

One goal while I was writing this book was to determine the kinds of changes that take place at different levels of biological organization as a consequence of selection on a colony-level phenotype. Another was to map mechanisms across levels and construct a cross-level metamap of related traits and their effects. In this chapter, I reconstruct what I have presented in Chapters 5–8 and present such a map.

9.1 Loading Algorithms

The regulation of pollen hoarding involves the local loading decisions (algorithms) of foragers as they visit flowers and recruitment of foragers for resources. Loading algorithms are influenced by within-hive stimuli, such as the quantity of stored pollen and brood pheromone. Keith Waddington showed that pollen quality also affects recruitment behavior for pollen, although we don't know how it affects pollen collection at flowers. However, the collection of pollen and nectar are intimately related. Most bees collect both. Pollen loads are less dense than nectar and pollen loads of weight equivalent to nectar would be very large protrusions from the hind legs of foragers. The total weight of the loads and probably the aerodynamics of large pollen loads place loading limits on bees. A totally loaded bee can carry about 30 mg of pollen if she does not carry nectar and about 60 mg of nectar if she does not carry pollen (Figure 2.12). How, then, does a forager "decide" how much of each she will collect on an individual foraging trip? Is she loading according to stimuli associated with nectar collecting, pollen collecting, or both? Adam Siegel asked these questions as part of his doctoral research in my laboratory at Arizona State University.

It is very easy to determine the loading of nectar. Artificial feeders can be constructed that mimic the shape, color, odor, and rate of delivery of flowers. However, it is difficult to study pollen loading because we do not have an artificial flower that delivers pollen like a flower. Instead, to study pollen collection, we take pollen that has been collected by bees and put it in open dishes. Bees will collect the pollen in the dishes, but they do so by wallowing in it and then hovering over the dish while they brush the pollen off their bodies and pack it into their pollen baskets (Figure 9.1). Then they wallow and pack again and again until their pollen baskets are completely full. I call them pollen "pigs." This is very unnatural. In addition, when we offer pollen and sugar-solution feeders at the same time in a cage where bees are restricted to foraging on them, bees normally will collect

Figure 9.1. Bees collecting pollen from a dish. Photo by Kim Fondrk.

only sugar solution or only pollen, even when the feeders are side by side.

Adam Siegel studied nectar loading. His hypothesis was that the ovary affects sucrose sensitivity, which affects nectar loading, which in turn affects pollen load size. He started with the knowledge that (1) nectar loading negatively affects pollen loading, (2) sucrose responsiveness correlates with nectar and pollen loads, and (3) ovary size affects sucrose responsiveness and foraging load. He set up a foraging cage and had bees forage at an artificial feeder that mimicked a flower and controlled the rate of delivery of sucrose solution. He offered bees solutions of different concentrations, determined the quantity of solution they collected and their responsiveness to sucrose using the proboscis extension response (PER) assay, and dissected their ovaries to determine the number of ovarioles. He found that sucrose responsiveness correlated with the concentration of solution the bees would collect, the concentration affected the amount of solution collected, and the number of ovarioles correlated with the relationship between the amount of solution collected and the concentration. The ovary apparently tuned the loading algorithm.

9.2 Heritability of the Pollen-Hoarding Syndrome

How much of the phenotypic architecture of the traits associated with pollen hoarding do we explain by the genetic architecture? The answer is "a lot and very little." Each generation, we measured the amount of pollen stored by high- and low-strain colonies. Each generation, we calculated the proportion of the total variance of the pooled colonies (high- and low-strain colonies together) that was a consequence of being from the high or low strain. This is a measure of the genetic separation of the strains (heritability or genetic determination; see Section 5.2.1). The average proportion was 0.41; in other words, the genetic differences between the strains determined 41 percent of the total variance (Figure 9.2). This value is very high and shows a very strong response to our selection.

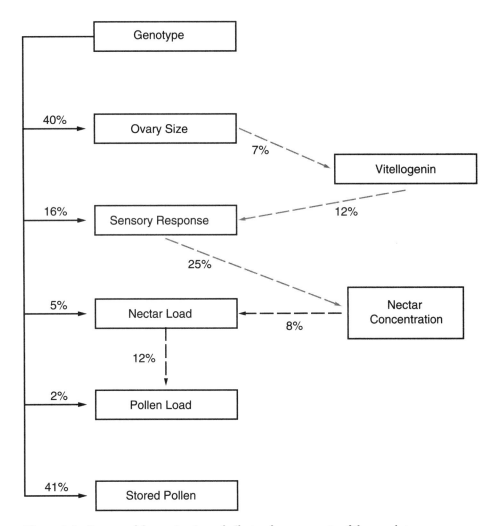

Figure 9.2. Degree of determination of effects of components of the regulatory architecture of pollen hoarding depicted as stored pollen in the bottom box. Genotype effects are shown down the left margin for each of the traits on the left. Arrows show the directionality of the effects. Solid-line relationships were measured in high- and low-strain bees; dark dashed lines were measured in high- and low-strain bees and wild-type bees; light dashed lines were measured in wild-type bees only. Reprinted from *Hormones, Brain and Behavior*, 2nd edition, Vol. 2. Amdam GV, Ihle DE, Page RE, "Regulation of Honeybee Worker (*Apis mellifera*) Life Histories by Vitellogenin," in: Donald W. Pfaff, Arthur P. Arnold, Anne M. Etgen, Susan E. Fahrbach and Robert T. Rubin, editors. San Diego: Academic Press, 1003–1025, Fig. 6 (2009), with permission from Elsevier.

We looked at ovariole number, an individual anatomical trait. Strain explained about 40 percent of the variance, similar to the social trait of pollen hoarding. We also looked at individual foraging traits. For example, we performed common-garden experiments (Figure 4.5) where we put marked high- and low-strain bees in common wild-type colonies, collected returning foragers, measured their loads, and partitioned the variance of the total, combined sample. The amount of variance explained by strain varied between 2 percent and 16 percent.

The combination of the phenotypic architecture and the reproductive-ground-plan hypothesis suggests a hypothesis for the causal links between the pollen-hoarding genetic architecture and foraging behavior. One can debate the degree to which we have been able to distinguish between direct causal pathways and latent unknown variables affecting multiple factors, but this is our working hypothesis. The pollen-hoarding quantitative trait loci (QTLs) affect ovary development, as is demonstrated by differences in ovariole number between strains, which in turn affects vitellogenin levels in developing adult bees, which in turn affects response thresholds to sucrose throughout life. Sucrose responsiveness affects the concentration of nectar collected, and the amount of collected nectar affects the amount of pollen collected. I constructed the correlative relationships between the regulatory elements in the chain of hypothesized causal factors (Figure 9.2). Some patterns are obvious.

1. Genetic differences between highs and lows explain much more of the phenotypic variance at the social phenotype level (stored pollen) than at the individual foraging level. This is the opposite of what I expected. I expected that the further we got from the more direct action of the genes involved, the less genetic effect we would see. Stored pollen is a consequence of many interacting features of the social and foraging environment.

2. The genetic effects on ovariole number are strong, about equal to the effects on stored pollen. I am not surprised, because I think that ovary development is closer to the direct action of the genes than the foraging behavior that they influence.

3. Genetic effects on the sensory-response system are stronger than on foraging behavior. This was expected because the sensory responses lie between the action of genes and the behavior of the bees.

4. The correlations between the causal links are mostly weak. The correlation between ovariole number and vitellogenin is not expected to be strong because Vg titers are dynamic and, therefore, are subject to large fluctuations depending on when they are measured during the life history of the bee. The model presented in Figure 7.5 suggests that Vg in some way sets the system during a window of time soon after bees become adults; afterward, the response system is independent of current blood titers. The effect of Vg must work like this because foragers have very little circulating Vg. The correlation between Vg and sucrose response suffers the same difficulty, with the added problem that sucrose responsiveness changes with age, perhaps because of declining Vg and increasing JH, and nutritional state. The relatively high correlation between sucrose response and the concentration of nectar collected in the field is, therefore, surprising if one considers how variable and unpredictable the foraging environment can be on an hour-to-hour basis. The relatively low correlations of nectar load to nectar concentration and pollen load to nectar load are not surprising, given the changing and unpredictable foraging environment.

9.3 Social Regulation of Pollen Hoarding

How can genotype (strain) explain so much of the phenotypic variation at the colony level (stored pollen) and so little at the individual foraging level? The answer, of course, is captured by Maurice Maeterlinck's spirit of the hive. It is a consequence of the stimulus-response relationships of thousands of individuals who alter the stimulus environment through their individual actions (Figure 9.3). Foragers collect loads of pollen and nectar according to a loading algorithm that is determined by their sensory-response system and stimuli from the nest and foraging environment. The pollen they collect contributes to the pollen intake of the colony, which gets stored as pollen, consumed, and converted into brood. Stored pollen is a negative pollen-foraging stimulus, while larvae provide

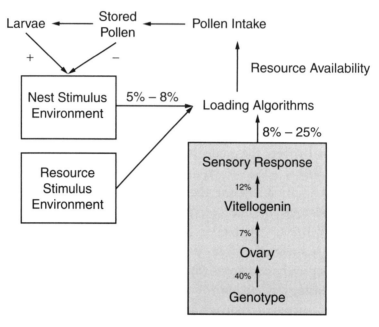

Figure 9.3. The regulatory architecture of pollen hoarding. The percentage explained between levels, when known, is shown next to each arrow. Arrows show directionality of relationships between components. The box on the right shows relationships "inside" foragers. Reprinted from *Hormones, Brain and Behavior*, 2nd edition, Vol. 2. Amdam GV, Ihle DE, Page RE, "Regulation of Honeybee Worker (*Apis mellifera*) Life Histories by Vitellogenin," in: Donald W. Pfaff, Arthur P. Arnold, Anne M. Etgen, Susan E. Fahrbach and Robert T. Rubin, editors. San Diego: Academic Press, 1003–1025, Fig. 7 (2009), with permission from Elsevier.

a positive stimulus. For any given amount of larvae in a colony, stored pollen is the regulatory target of the foragers. Each forager has its own target, which is based in part on its genotype. There is no single target for the colony. The resulting amount of stored pollen is a consequence of the distribution of targets for all foragers, the consumption rate, and the availability of pollen in the field. Individual foragers will continue to collect pollen until the amount stored exceeds the targets of all individuals. However, an equilibrium will be reached at the colony level where those that continue to forage for pollen or bias their loads toward pollen are offset by the rate of consumption. This is similar in principle to the model shown in panel D of Figure 2.4.

I am frequently asked, "But how much of the loading algorithm is determined by colony need and how much by the genotype of the worker?" Genotype appears not to contribute much to individual load size, only about 8 to 25 percent, usually closer to 8 percent, depending on the individual experiment. We conducted an experiment where we varied the quantities of stored pollen and brood to maximize differences in colony need, an empirical approach similar to the ensemble modeling discussed in Chapter 2. Some colonies had large quantities of stored pollen and very few larvae (low need), while others had very little stored pollen and large numbers of larvae (high need). Over this extreme range of colony stimuli (need), colony stimuli explained only 5 to 8 percent of the variance in pollen and nectar loading. It explained no more than genotype. What explains the rest? It must be the chance daily and seasonal availability of pollen and nectar.

Throughout this book, I have decomposed the regulatory architecture at different levels of organization spanning genes and complex behavioral interactions. I have ended this chapter with a regulatory metamodel. However, the model is a very thin soup made of what we know now about the architecture of the genome, gene function and regulation, and hormonal control systems. In time, members of the honey bee research village will undoubtedly add more ingredients to the kettle and make a more complete model.

Suggested Reading

Amdam, G. V., Ihle, K. E., and Page, R. E. 2009. Regulation of honeybee worker *(Apis mellifera)* life histories by vitellogenin. In *Hormones, Brain and Behavior,* 2nd ed., vol. 2, ed. D. W. Pfaff, A. P. Arnold, A. M. Etgen, S. E. Fahrbach, et al. San Diego: Academic Press, pp. 1003–1025.

Dreller, C., Page, R. E., and Fondrk, M. K. 1999. Regulation of pollen foraging in honeybee colonies: Effects of young brood, stored pollen, and empty space. *Behav. Ecol. Sociobiol.* 45:227–233.

Dreller, C. R., and Tarpy, D. R. 2000. Perception of the pollen need by foragers in a honeybee colony. *Anim. Behav.* 59:91–96.

Pankiw, T., and Page, R. E. 2000. Response thresholds to sucrose predict foraging division of labor in honeybees. *Behav. Ecol. Sociobiol.* 47:265–267.

Pankiw, T., and Page, R. E. 2001. Genotype and colony environment affect honey bee (*Apis mellifera* L.) development and foraging behavior. *Behav. Ecol. Sociobiol.* 51:87–94.

Pankiw, T., Page, R. E., and Fondrk, M. K. 1998. Brood pheromone stimulates pollen foraging in honey bees *(Apis mellifera). Behav. Ecol. Sociobiol.* 44:193–198.

Siegel, A. J., Freedman, C., and Page, R. E. 2012. Ovarian control of nectar collection in the honey bee *(Apis mellifera). PloS ONE* 7:e33465. doi:10.1371/journal.pone.0033465.

Tsuruda, J. M., Amdam, G. V., and Page, R. E. 2008. Sensory response system of social behavior tied to female reproductive traits. *PLoS ONE* 3:e3397. doi:10.1371/journal.pone.0003397.

Waddington, K. D., Nelson, C. M., and Page, R. E. 1998. Effects of pollen quality and genotype on the dance of foraging honey bees. *Anim. Behav.* 56:35–39.

—10—

A Crowd of Bees

Charles Darwin marveled at how honey bees can construct a wax comb even though each bee has only her instincts and limited information about her own small local environment. The end product is, however, in his mind nearly perfect. Maurice Maeterlinck, 50 years later, pondered the source of the organized social behavior and proposed a mystical "spirit of the hive," wondering where it resides. In this book, I have shown that the stimulus-response relationships of individual bees, coupled with the effects of the behavior of individuals on the environment they share with nestmates, provide the basic mechanism for social behavior. I call it the spirit of the hive, solving one part of Maeterlinck's puzzle. Complex social organization evolves by tuning the stimulus-response relationships of individuals.

Social behavior is woven from the fabric of ancient genetic and physiological regulatory systems derived from solitary insects. When we were selecting for a social phenotype, the amount of stored pollen, we affected a network of genetic and phenotypic traits that are linked within and among levels (Figure 6.5). Selection for pollen hoarding resulted in changes in the frequencies of alternative genes or genetic regulatory systems, demonstrated by mapped QTLs. They in turn affected fundamental larval developmental processes, which resulted in changes in reproductive anatomy and physiology and probably changes in the evolutionarily ancient insulin-insulin-like-signaling pathways that affect development and behavior. The ultimate target, however, is the brain. The complex web of interacting genes and regulatory pathways ultimately modulates neurobiochemical systems that modulate sensory-response systems. However, I will leave the challenge of mapping the neurobiology of social behavior to others.

It is clear that it is not sufficient to look at any one level if one wants to understand the mechanisms and evolution of social behavior. However, looking at multiple levels at once requires different models than currently exist. The question is too complex for stone soup, but we don't want to make Mulligan stew. We want to be able to understand the necessary and sufficient components, even though they are complex. Many today are still looking for "a gene for trait X" that is different from "a gene for trait Y." Ernst Mayr, one of the architects of the modern evolutionary synthesis, referred to this kind of thinking as "beanbag genetics." Beanbag thinking is not different from the analogy of trying to understand what the beach looks like by examining it one grain of sand at a time. With the growing set of available genomic tools, it is tempting to treat the honey bee genome as a very large collection of bean bags and continue to look for the gene or the collection of genes responsible for some trait. Instead, we need to go up to 10,000 meters and study the beach. We need to understand the ecology of genes, where the genome and the physiology of the individual constitute the ecological background. We should also extend this approach to include the ecology of the environment, an equal partner during development, and recognize that phenotypic traits are influenced by networks of processes and are linked together mechanistically. The balloon metaphor used in Section 5.12 comes to mind.

This has been the story of more than 30 years of research. I have tried to link together topics seemingly as disparate as sex determination, mating behavior, variation in individual behavior, phenotypic and genetic architectures, development, reproductive anatomy and physiology, and social foraging. Finally, I return to the second part of Maeterlinck's puzzle. Where does the spirit of the hive reside? At least to some extent it is in the ovaries of "a crowd of bees working in a dark hive."

Acknowledgments

This book represents the efforts of a community of students, postdoctoral researchers, and colleagues spanning more than 30 years. Many are mentioned in this book, but I have too little room to recognize all of them fairly. But you all know who you are. Thank you for your hard work and support.

Jerry Downhower was my closest faculty colleague when I was at The Ohio State University. One day in 2008, I received a small package in the mail. It was from Jerry. I had not heard from him for a few years, since his retirement. I opened it and found a book he had purchased for $2.00 at a used-book sale. He told me that he thought about me when he saw it. The book was *The Life of the Bee* by Maurice Maeterlink. It became the inspiration for this book.

Jochen Erber made science fun for me again. I spent a year working in his lab on a Humboldt Research Prize in 1996. I was burned out and needed a change, and he provided me with an environment where research was interesting and challenging. We formed a close collaboration that lasted 15 years, until his retirement. The core of this book is about phenotypic architecture. We built most of that together, and without him the scaffold would be thin.

I reserve my greatest thanks and appreciation for Kim Fondrk. Kim became my technician in 1986 when I accepted my first faculty position at The Ohio State University. He moved with me to the University of California–Davis and then to Arizona State University. He retired from ASU but continues to support my research by maintaining the pollen-hoarding selection program at UC Davis. Kim is a very special person. He is a loyal technician, an excellent colleague, a full intellectual partner in all that we do, and a good friend. Truly, without him this book would be empty.

I also thank the team with the Visualization Laboratory at Arizona State University, Jacob Sahertian, Jacob Mayfield, and Sabine Deviche, for drawing, designing, and cleaning up many of the figures, and especially Margaret (Peggy) Coulombe for her help with editing, formatting, and coordinating. She was essential for bringing the book project to completion.

Much of this book was written while I was a fellow at the Wissenschaftskolleg zu Berlin (Institute for Advanced Studies, Berlin). Their continuing support of my scholarship effort is greatly appreciated. Parts of Chapter 1 and Chapter 2 of this book first appeared in my 2009 Ernst Mayr Lecture and were published with permission from the Berlin-Brandenburg Academy of Science. My connections with Germany have been very important for my research, and to me personally, since 1996. They were made possible by the Alexander von Humboldt Foundation, for which I am very grateful.

Index

Note: Page numbers followed by *f* and *t* indicate figures and tables.